Vegetable Diet

As Sanctioned by Medical Men, and by Experience in All Ages; Including a System of Vegetable Cookery

William A. Alcott

Alpha Editions

This edition published in 2024

ISBN : 9789362923837

Design and Setting By
Alpha Editions
www.alphaedis.com
Email - info@alphaedis.com

As per information held with us this book is in Public Domain. This book is a reproduction of an important historical work. Alpha Editions uses the best technology to reproduce historical work in the same manner it was first published to preserve its original nature. Any marks or number seen are left intentionally to preserve its true form.

Contents

PREFACE .... - 1 -
CHAPTER I. .... - 3 -
CHAPTER II. .... - 8 -
CHAPTER III. .... - 32 -
CHAPTER IV. .... - 38 -
CHAPTER V. .... - 55 -
CHAPTER VI. .... - 107 -
CHAPTER VII. .... - 136 -
CHAPTER VIII. .... - 144 -
OUTLINES OF A NEW SYSTEM OF FOOD AND COOKERY. .... - 177 -

PREFACE

The following volume embraces the testimony, direct or indirect, of more than a HUNDRED individuals—besides that of societies and communities—on the subject of vegetable diet. Most of this one hundred persons are, or were, persons of considerable distinction in society; and more than FIFTY of them were either medical men, or such as have made physiology, hygiene, anatomy, pathology, medicine, or surgery a leading or favorite study.

As I have written other works besides this—especially the "Young House-Keeper"—which treat, more or less, of diet, it may possibly be objected, that I sometimes repeat the same idea. But how is it to be avoided? In writing for various classes of the community, and presenting my views in various connections and aspects, it is almost necessary to do so. Writers on theology, or education, or any other important topic, do the same—probably to a far greater extent, in many instances, than I have yet done. I repeat no idea for the *sake* of repeating it. Not a word is inserted but what seems to me necessary, in order that I may be intelligible. Moreover, like the preacher of truth on many other subjects, it is not so much my object to produce something new in every paragraph, as to explain, illustrate, and enforce what is already known.

It may also be thought that I make too many books. But, as I do not claim to be so much an originator of *new* things as an instrument for diffusing the *old*, it will not be expected that I should be twenty years on a volume, like Bishop Butler. I had, however, been collecting my stock of materials for this and other works—published or unpublished—more than twenty-five years. Besides, it might be safely and truly said that the study and reading and writing, in the preparation of this volume, the "House I Live In," and the "Young House-Keeper," have consumed at least three of the best years of my life, at fourteen or fifteen hours a day. Several of my other works, as the "Young Mother," the "Mother's Medical Guide," and the "Young Wife," have also been the fruit of years of toil and investigation and observation, of which those who think only of the labor of merely *writing them out*, know nothing. Even the "Mother in her Family"—at least some parts of it—though in general a lighter work, has been the result of much care and labor. The circumstance of publishing several books at the same, or nearly the same time, has little or nothing to do with their preparation.

When I commenced putting together the materials of this little treatise on diet—thirteen years ago—it was my intention simply to show the SAFETY of a vegetable and fruit diet, both for those who are afflicted with many forms of chronic disease, and for the healthy. But I soon became convinced that I ought to go farther, and show its SUPERIORITY over every other. This I have attempted to do—with what success, the reader must and will judge for himself.

I have said, it was not my original intention to prove a vegetable and fruit diet to be any thing more than *safe*. But I wish not to be understood as entertaining, even at that time, any doubts in regard to the superiority of such a diet: the only questions with me were, Whether the public mind was ready to hear and weigh the proofs, and whether this volume was the place in which to present them. Both these questions, however, as I went on, were settled, in the affirmative. I believed—and still believe—that the public mind, in this country, is prepared for the free discussion of all topics—provided they are discussed candidly—which have a manifest bearing on the well-being of man; and I have governed myself accordingly.

An apology may be necessary for retaining, unexplained, a few medical terms. But I did not feel at liberty to change them, in the correspondence of Dr. North, for more popular language; and, having retained them thus far, it did not seem desirable to explain them elsewhere. Nor was I willing to deface the pages of the work with explanatory notes. The fact is, the technical terms alluded to, are, after all, very few in number, and may be generally understood by the connection in which they appear.

THE AUTHOR.
WEST NEWTON Mass.

CHAPTER I.

ORIGIN OF THIS WORK.

Experience of the Author, and his Studies.—Pamphlet in 1832.—Prize Question of the Boylston Medical Committee.—Collection of Materials for an Essay.—Dr. North.—His Letter and Questions.—Results.

Twenty-three years ago, the present season, I was in the first stage of tuberculous consumption, and evidently advancing rapidly to the second. The most judicious physicians were consulted, and their advice at length followed. I commenced the practice of medicine, traveling chiefly on horseback; and, though unable to do but little at first, I soon gained strength enough to perform a moderate business, and to combine with it a little gardening and farming. At the time, or nearly at the time, of commencing the practice of medicine, I laid aside my feather bed, and slept on straw; and in December, of the same year, I abandoned spirits, and most kinds of stimulating food. It was not, however, until nineteen years ago, the present season, that I abandoned all drinks but water, and all flesh, fish, and other highly stimulating and concentrated aliments, and confined myself to a diet of milk, fruits, and vegetables.

In the meantime, the duties of my profession, and the nature of my studies led me to prosecute, more diligently than ever, a subject which I had been studying, more or less, from my very childhood—the laws of Human Health. Among other things, I collected facts on this subject from books which came in my way; so that when I went to Boston, in January, 1832, I had already obtained, from various writers, on materia medica, physiology, disease, and dietetics, quite a large parcel. The results of my reflections on these, and of my own observation and experience, were, in part—but in part only—developed in July, of the same year, in an anonymous pamphlet, entitled, "Rational View of the Spasmodic Cholera;" published by Messrs. Clapp & Hull, of Boston.

In the summer of 1833, the Boylston Medical Committee of Harvard University offered a prize of fifty dollars, or a gold medal of that value, to the author of the best dissertation on the following question: "What diet can be selected which will ensure the greatest health and strength to the laborer in the climate of New England—quality and quantity, and the time and manner of taking it, to be considered?"

At first, I had thoughts of attempting an essay on the subject; for it seemed to me an important one. Circumstances, however, did not permit me to prosecute the undertaking; though I was excited by the question of the

Boylston Medical Committee to renewed efforts to increase my stock of information and of facts.

In 1834, I accidentally learned that Dr. Milo L. North, a distinguished practitioner of medicine in Hartford, Connecticut, was pursuing a course of inquiry not unlike my own, and collecting facts and materials for a similar purpose. In correspondence with Dr. North, a proposition was made to unite our stock of materials; but nothing for the present was actually done. However, I agreed to furnish Dr. North with a statement of my own experience, and such other important facts as came within the range of my own observations; and a statement of my experience was subsequently intrusted to his care, as will be seen in its place, in the body of this work.

In February, 1835, Dr. North, in the prosecution of his efforts, addressed the following circular, or LETTER and QUESTIONS, to the editor of the Boston Medical and Surgical Journal, which were accordingly inserted in a subsequent number of that work. They were also published in the American Journal of Medical Science, of Philadelphia, and copied into numerous papers, so that they were pretty generally circulated throughout our country.

"To the Editor of the Boston Medical and Surgical Journal.

"SIR,—Reports not unfrequently reach us of certain individuals who have fallen victims to a prescribed course of regimen. Those persons are said, by gentlemen who are entitled to the fullest confidence, to have pertinaciously followed the course, till they reached a point of reduction from which there was no recovery. If these are facts, they ought to be collected and published. And I beg leave, through your Journal, to request my medical brethren, if they have been called to advise in such cases, that they will have the kindness to answer, briefly, the following interrogatories, by mail, as early as convenient.

"Should the substance of their replies ever be embodied in a small volume, they will not only receive a copy and the thanks of the author, but will have the pleasure to know they are assisting in the settlement of a question of great interest to the country. If it should appear probable that their patient was laboring under a decline at the commencement of the change of diet, this ought, in candor, to be fully disclosed.

"It will be perceived, by the tenor of the questions, that they are designed to embrace not only unfortunate results of a change of diet, but such as are favorable. There are, in our community, considerable numbers who have entirely excluded animal food from their diet. It is exceedingly desirable that the results of such experiments, so difficult to be found in this land of plenty, should be ascertained and thrown before the profession and the community. Will physicians, then, have the kindness, if they know of any persons in their vicinity who have excluded animal food from their diet for a year or over, to lend them this number of the Journal, and ask them to forward to Milo L. North, Hartford, Connecticut, as early as convenient, the result of this change of diet on their health and constitution, in accordance with the following inquiries?

"1. Was your bodily strength either increased or diminished by excluding all animal food from your diet?

"2. Were the animal sensations, connected with the process of digestion, more—or less agreeable?

"3. Was the mind clearer; and could it continue a laborious investigation longer than when you subsisted on mixed diet?

"4. What constitutional infirmities were aggravated or removed?

"5. Had you fewer colds or other febrile attacks—or the reverse?

"6. What length of time, the trial?

"7. Was the change to a vegetable diet, in your case, preceded by the use of an uncommon proportion of animal food, or of high seasoning, or of stimulants?

"8. Was this change accompanied by a substitution of cold water for tea and coffee, during the experiment?

"9. Is a vegetable diet more—or less aperient than mixed?

"10. Do you believe, from your experience, that the health of either laborers or students would be promoted by the exclusion of animal food from their diet?

"11. Have you selected, from your own observation, any articles in the vegetable kingdom, as particularly healthy, or otherwise?

"N.B.—Short answers to these inquiries are all that is necessary; and as a copy of the latter is retained by the writer, it will be sufficient to refer to them numerically, without the trouble of transcribing each question.

"HARTFORD, February 25, 1835."

This circular, or letter, drew forth numerous replies from various parts of the United States, and chiefly from medical men. In the meantime, the prize of the Boylston Medical Committee was awarded to Luther V. Bell, M.D., of Derry, New Hampshire, and was published in the Boston Medical and Surgical Journal, and elsewhere, and read with considerable interest.

In the year 1836, while many were waiting—some with a degree of impatience—to hear from Dr. North, his health so far failed him, that he concluded to relinquish, for the present, his inquiries; and, at his particular request, I consented to have the following card inserted in the Boston Medical and Surgical Journal:

"DR. NORTH, of Hartford, Connecticut, tenders his grateful acknowledgments to the numerous individuals, who were so kind as to forward to him a statement of the effects of vegetable diet on their own persons, in reply to some specific inquiries inserted in the Boston Medical and Surgical Journal of March 11, 1835, and in the Philadelphia Journal of the same year. Although many months elapsed before the answers were all received, yet the writer is fully aware that these communications ought to have been published before this. His apology is a prolonged state of ill health, which has now become so serious as to threaten to drive him to a southern climate for the winter. In this exigency, he has solicited Dr. W. A. Alcott, of Boston, to receive the papers and give them to the public as soon as his numerous engagements will permit. This arrangement will doubtless be fully satisfactory, both to the writers of the communications and to the public.

"HARTFORD, November 4, 1836."

Various circumstances, beyond my control, united to defer the publication of the contemplated work to the year 1838. It is hoped, however, that nothing was lost by delay. It gave further opportunity for reflection, as well as for observation and experiment; and if the work is of any value at all to the community, it owes much of that value to the fact that what the public may be disposed to regard as unnecessary, afforded another year for investigation. Not that any new discoveries were made in that time, but I was, at least, enabled to verify and confirm my former conclusions, and to review, more carefully than ever, the whole argument. It is hoped that the work will at least serve as a pioneer to a more extensive as well as more scientific volume, by some individual who is better able to do the subject justice.

It will be my object to present the facts and arguments of the following volume, not in a distorted or one-sided manner, but according to truth. I have no private interests to subserve, which would lead me to suppress, or falsely color, or exaggerate. If vegetable food is not preferable to animal, I certainly do not wish to have it so regarded. This profession of a sincere desire to know and teach the truth may be an apology for placing the letters in the order in which they appear—which certainly is such as to give no unfair advantages to those who believe in the superiority of the vegetable system—and for the faithfulness with which their whole contents, whether favoring one side or other of the argument, have been transcribed.

The title of the work requires a word of explanation. It is not intended, or even intimated, that there are no facts here but what rest on medical authority; but rather, that the work originated with the medical profession, and contains, for the most part, testimony which is exclusively medical—either given by medical men, or under their sanction. In fact, though designed chiefly for popular reading, it is in a good degree a medical work; and will probably stand or fall, according to the sentence of approbation or disapprobation which shall be pronounced by the medical profession.

The following chapter will contain the letters addressed to Dr. North. They are inserted, with a single exception, in the precise order of their date. The first, however, does not appear to have been elicited by Dr. North's circular; but rather by a request in some previous letter. It will be observed that several of the letters include more than one case or experiment; and a few of them many. Thus the whole series embraces, at the least calculation, from thirty to forty experiments.

The replies of nearly every individual are numbered to correspond with the questions, as suggested by Dr. North; so that, if there should remain a doubt, in any case, in regard to the precise point referred to by the writer of the letter, the reader has only to turn to the circular in the present chapter, and read the question there, which corresponds to the number of the doubtful one. Thus, for example, the various replies marked 6, refer to the length or duration of the experiment or experiments which had been made; and those marked 9, to the aperient effects of a diet exclusively vegetable. And so of all the rest.

CHAPTER II.

LETTERS TO DR. NORTH.

Letter of Dr. Parmly.—Dr. W. A. Alcott.—Dr. D. S. Wright.—Dr. H. N. Preston.—Dr. H. A. Barrows.—Dr. Caleb Bannister.—Dr. Lyman Tenny.—Dr. J. M. B. Harden.—Joseph Ricketson, Esq.—Joseph Congdon, Esq.—George W. Baker, Esq.—John Howland, Jr., Esq.—Dr. Wm. H. Webster.—Josiah Bennet, Esq.—Wm. Vincent, Esq.—Dr. Geo. H. Perry.—Dr. L. W. Sherman.

LETTER I.—FROM DR. PARMLY, DENTIST.

To Dr. North.

MY DEAR SIR,—For two years past, I have abstained from the use of all the diffusible stimulants, using no animal food, either flesh, fish, or fowl; nor any alcoholic or vinous spirits; no form of ale, beer, or porter; no cider, tea, or coffee; but using milk and water as my only liquid aliment, and feeding sparingly, or rather, moderately, upon farinaceous food, vegetables, and fruit, seasoned with unmelted butter, slightly boiled eggs, and sugar or molasses; with no condiment but common salt.

I adopted this regimen in company with several friends, male and female, some of whom had been afflicted either with dyspepsia or some other chronic malady. In every instance within the circle of my acquaintance, the *symptoms* of disease disappeared before this system of diet; and I have every reason to believe that the disease itself was wholly or in part eradicated.

In answer to your inquiry, whether I ascribe the cure, in the cases alleged, to the abstinence from animal food or from stimulating drinks, or from both, I cannot but give it as my confident opinion that the result is to be attributed to a general abandonment of the *diffusive stimuli*, under every shape and form.

An increase of flesh was one of the earliest effects of the *anti-stimulating* regimen, in those cures in which the system was in low condition. The animal spirits became more cheerful, buoyant, and uniformly pleasurable. Mental and bodily labor was endured with much less fatigue, and both intellectual and corporeal exertion was more vigorous and efficient.

In the language of Addison, this system of ultra temperance has had the happy effect of "filling the mind with inward joy, and spreading delight through all its faculties."

But, although I have thus made the experiment of abstaining wholly from the use of liquid and solid stimulants, and from every form of animal food, I am not fully convinced that it should be deemed improper, on any account, to use the more slightly stimulating forms of animal food. Perhaps fish and fowl, with the exception of ducks and geese, turtle and lobster, may be taken without detriment, in moderate quantities. And I regard good mutton as being the lightest, and, at the same time, the most nutritious of all meats, and as producing less inconvenience than any other kind, where the energies of the stomach are enfeebled. And yet there are unquestionably many constitutions which would be benefited by living, as I and others have done, on purely vegetable diet and ripe fruits.

In relation to many of the grosser kinds of animal food, all alcoholic spirits, all distilled and fermented liquors, tea and coffee, opium and tobacco,—I feel confident in pronouncing them not only useless, but noxious to the animal machine.

Yours, etc.,

ELEAZER PARMLY

NEW YORK, January 31, 1835.

LETTER II—FROM DR. W. A. ALCOTT.

BOSTON, December 19, 1834.

DEAR SIR,—I received your communication, and hasten to reply to as many of your inquiries as I can. Allow me to take them up in the very order in which you have presented them.

Answer to question 1. I was bred to a very active life, from my earliest childhood. This active course was continued till about the time of my leaving off the use of flesh and fish; since which period my habits have, unfortunately, been more sedentary. I think my muscular strength is somewhat less now than it was before I omitted flesh meat, but in what proportion I am unable to say; for indeed it varies greatly. When more exercise is used, my strength increases—sometimes almost immediately; when less exercise is used, my strength again diminishes, but not so rapidly. These last circumstances indicate a more direct connection between my loss of muscular strength and my neglect of exercise than between the former and my food.

2. Rather more agreeable; unless I use too large a quantity of food; to which however I am rather more inclined than formerly, as my appetite is keener, and food relishes far better. A sedentary life, moreover, as I am well satisfied, tends to bring my moral powers into subjection to the physical.

3. My mind has been clearer, since I commenced the experiment to which you allude, than before; but I doubt whether I can better endure a "laborious investigation." A little rest or exercise, perhaps less than formerly, restores

vigor. I am sometimes tempted to *break my day into two*, by sleeping at noon. But I am not so apt to be cloyed with study, or reflection, as formerly.

4. Several. 1. An eruptive complaint, sometimes, at one period of my life, very severe. 2. Irritation of the lungs; probably, indeed most certainly, incipient phthisis. 3. Rheumatic attacks, though they had never been very severe.

The eruptive disease, however, and the rheumatic attacks, are not wholly removed; but they are greatly diminished. The irritation at the lungs has nearly left me. This is the more remarkable from the fact that I have been, during almost the whole period of my experiment, in or about Boston. I was formerly somewhat subject to palpitations; these are now less frequent. I am also less exposed to epidemics. Formerly, like other scrofulous persons, I had nearly all that appeared; now I have very few.

You will observe that I merely state the facts, without affirming, positively, that my change of diet has been the cause, though I am quite of opinion that this has not been without its influence. Mental quiet and total abstinence from all drinks but water, may also have had much influence, as well as other causes.

5. Very few colds. Last winter I had a violent inflammation of the ear, which was attended with some fever; but abstinence and emollient applications soon restored me. In July last, I had a severe attack of diarrhœa unattended with much fever, which I attributed to drinking too much water impregnated with earthy salts, and to which I had been unaccustomed. When I have a cold, of late, it affects, principally, the nasal membrane; and, if I practice abstinence, soon disappears. In this respect, more than in any other, I am confident that since I commenced the use of a vegetable diet I have been a very great gainer.

6. The experiment was fully begun four years ago last summer; though I had been making great changes in my physical habits for four years before. For about three years, I used neither flesh nor fish, nor even eggs more than two or three times a year. The only animal food I used was milk; and for some long periods, not even that. But at the end of three years I ate a very small quantity of flesh meat once a day, for three or four weeks, and then laid it aside. This was in the time of the cholera. The only effect I perceived from its use was a slight increase of peristaltic action. In March last, I used a little dried fish once or twice a day, for a few days; but with no peculiar effects. After my attack of diarrhœa, in July last, I used a little flesh several times; but for some months past I have laid it aside entirely, with no intention of resuming it. Nothing peculiar was observed, as to its effects, during the last autumn.

7. I never used a large proportion of animal food, except milk, since I was a child; but I have been in the habit, at various periods of my life, of drinking considerable cider. For some months before I laid aside flesh and fish, I had been accustomed to the use of more animal food than usual, but less cider; though, for a part of the time, I made up the deficiency of cider with ale and coffee. For several months previous to the beginning of the experiment, I had drank nothing but water.

8. Rather less. But here, again, I fear I am in danger of attributing to one cause what is the effect of another. My neglect of exercise may be more in fault than the rice and bread and milk which I use. Still I must think that vegetable food is, in my own case, less aperient than animal.

9. In regard to students, my reply is, Yes, most certainly. So I think in regard to laborers, were they trained to it. But how far *early habits* may create a demand for the continuance of animal food through life, I am quite at a loss for an opinion. Were I a hard laborer, I should use no animal food. When I travel on foot forty or fifty miles a day, I use vegetable food, and in less than the usual quantity. This I used to do before I commenced my experiment.

10. I use bread made of unbolted wheat meal, in moderate quantity, when I can get it; plain Indian cakes once a day; milk once a day; rice once a day. My plan is to use as few things as possible at the same meal, but to have considerable variety at different meals. I use no new bread or pastry, no cheese, and but little butter; and very little fruit, except apples in moderate quantity.

11. The answer to this question, though I think it would be important and interesting, with many other particulars, I must defer for the present. The experiments of Dr. F., a young man in this neighborhood, and of several other individuals, would, I know be in point; but I have not at my command the time necessary to present them.

LETTER III.—FROM DR. D. S. WRIGHT.

WHITEHALL, Washington Co., N. Y., March 17, 1835.

DEAR SIR,—I noticed a communication from you in the Boston Medical and Surgical Journal of the 5th instant, in which you signify a wish to collect facts in relation to the effects of a vegetable diet upon the human system, etc. I submit for your consideration my own experience; premising, however, that I am a practicing physician in this place—am thirty-three years old—of a sanguine, bilious temperament—have from youth up usually enjoyed good health—am not generally subject to fevers, etc.

I made a radical change in my diet three years ago this present month, from a mixed course of animal and vegetable food, to a strictly vegetable diet, on which I subsisted pretty uniformly for the most part of one year. I renewed it again about ten moths ago.

My reasons for adopting it were: 1st. I had experienced the beneficial effects of it for several years before, during the warm weather, in obviating a dull cephalalgic pain, and oppression in the epigastrium. 2dly. I had recently left the salubrious atmosphere of the mountains in Essex county, in this state, for this place of *musquitoes* and *miasmata*. 3dly, and prominently. I had frequent exposures to the variolous infection, and I had a *dreadful* apprehension that I might have an attack of the varioloid, as at that time I had never experimentally tried the protective powers of the vaccine virus, and had *too* little confidence in those who recommended its prophylactic powers. The results I submit you, in reply to your interrogatories.

1. I think each time I tried living on vegetable food exclusively, that for the first month I could not endure fatigue *as well*. Afterward I could.

2. The digestive organs were always more agreeably excited.

3. The mind uniformly clearer, and could endure laborious investigations longer, and with less effort.

4. I am constitutionally healthy and robust.

5. I believe I have more colds, principally seated on the mucous membranes of the lungs, fauces, and cavities of the head. (I do not, however, attribute it to diet.)

6. The first trial was one year. I am now ten months on the same plan, and shall continue it.

7. I never used a large quantity of animal food or stimulants, of any description.

8. I have for several years used tea and coffee, usually once a day—believe them healthy.

9. Vegetable diet is less aperient than a mixed diet, if we except *Indian corn*.

10. I do not think that common laborers, in health, could do as well without animal food; but I think students might.

11. I have selected *potatoes*, when *baked* or *roasted*, and all articles of food usually prepared from *Indian meal*, as the most healthy articles on which I subsist; particularly the latter, whose aperient and nutritive qualities render it, in my estimation, an invaluable article for common use.

Yours, etc.,

D. S. WRIGHT.

LETTER IV.—FROM DR. H. N. PRESTON.[1]

PLYMOUTH, Mass., March 26, 1835.

DEAR SIR,—When I observed your questions in the Boston Medical and Surgical Journal, of the 11th of March, I determined to give you personal experience, in reply to your valuable queries.

In the spring of 1832, while engaged in more than usual professional labor, I began to suffer from indigestion, which gradually increased, unabated by any medicinal or dietetic course, until I was reduced to the very confines of the grave. The disease became complicated, for a time, with chronic bronchitis. I would remark, that, at the time of my commencing a severe course of diet, I was able to attend to my practice daily.

In answer to your inquiries, I would say to the 1st—very much diminished, and rapidly.

2. Rather less; distinct local uneasiness—less disposition to drowsiness; but decidedly more troubled with cardialgia, and eructations.

3. I think not.

4. My disease was decidedly increased; as cough, headache, and emaciation; and being of a scrofulous diathesis, was lessening my prospect of eventual recovery.

5. My febrile attacks increased with my increased debility.

6. Almost four months; when I became convinced death would be the result, unless I altered my course.

7. I had taken animal food moderately, morning and noon—very little high seasoning—no stimulants, except tea and coffee. The latter was my favorite beverage; and I usually drank two cups with my breakfast and dinner, and black tea with my supper.

8. I drank but one cup of weak coffee with my breakfast, none with dinner, and generally a cup of milk and water with supper.

9. With me *much less aperient*; indeed, costiveness became a very serious and distressing accompaniment.

10. From somewhat extensive observation, for the last seven years, I should say, of laborers never; students seldom.

11. Among dyspeptics, potatoes nearly boiled, then mashed together, rolled into balls, and laid over hot coals, until a second time cooked, as easy as any vegetable. If any of the luxuries of the table have been noticed as particularly

injurious, it has been cranberries, prepared in any form, as stewed in sauce, tarts, pies, etc.

Crude as these answers are, they are at your service; and I am prompted to give them from the fact, that very few persons, I presume, have been so far reduced as myself, with dyspepsia and its concomitants. In fact, I was pronounced, by some of the most scientific physicians of Boston, as past all prospect of cure, or even much relief, from medicine, diet, or regimen. My attention has naturally been turned with anxious solicitude to the subject of diet, in all its forms. Since my unexpected restoration to health, my opportunities for observation among dyspeptics have been much enlarged; and I most unhesitatingly say, that my success is much more encouraging, in the management of such cases, since pursuing a more liberal diet, than before. Plain animal diet, avoiding condiments and tea, using mucilaginous drink, as the Irish Moss, is preferable to "absolute diet,"—cases of decided chronic gastritis excepted.

Yours, etc.,

H. N. PRESTON.

LETTER V.—FROM DR. H. A. BARROWS.

PHILLIPS, Somerset Co., Me., April 28, 1835.

DEAR SIR,—I have a brother-in-law, who owes his life to abstinence from animal food, and strict adherence to the simplest vegetable diet. My own existence is prolonged, only (according to human probabilities) by entire abstinence from flesh-meat of every description, and feeding principally upon the coarsest farinacea.

Numberless other instances have come under my observation within the last three years, in which a strict adherence to a simple vegetable diet has done for the wretched invalid what the best medical treatment had utterly failed to do; and in no one instance have I known permanently injurious results to follow from this course, but in many instances have had to lament the want of firmness and decision, and a gradual return to the "*flesh-pots of Egypt.*"

With these views, I very cheerfully comply with your general invitation, on page 77, volume 12, of the Boston Medical and Surgical Journal. The answers to your interrogatories will apply to the case first referred to, to my own case, and to nearly every one which has occurred within my notice.

1. Increased, uniformly; and in nearly every instance, without even the usual debility consequent upon withdrawing the stimulus of animal food.

2. More agreeable in every instance.

3. Affirmative, *in toto.*

4. None aggravated, except flatulence in one or two instances. All the horrid train of dyspeptic symptoms uniformly mitigated, and obstinate constipation removed.

5. Fewer colds and febrile attacks.

6. Three years, with my brother; with myself, eighteen months partially, and three months wholly; the others, from one to six months.

7. Negative.

8. Cold water—my brother and myself; others, hot and cold water alternately.

9. More aperient,—no exceptions.

10. I believe the health of *students* would uniformly be promoted—and the days of the laborer, to say the least, would be lengthened.

11. I have; and that is, simple bread made of wheat meal, ground in corn-stones, and mixed up precisely as it comes from the mill—with the substitution of fine flour when the bowels become too active.

Yours, etc.,

HORACE A. BARROWS.

LETTER VI.—FROM DR. CALEB BANNISTER.

PHELPS, N. Y., May 4, 1835.

SIR,—My age is fifty-three. My ancestors had all melted away with hereditary consumption. At the age of twenty, I began to be afflicted with pain in different parts of the thorax, and other premonitory symptoms of phthisis pulmonalis. Soon after this, my mother and eldest sister died with the disease. For myself, having a severe attack of ague and fever, all my consumptive symptoms became greatly aggravated; the pain was shifting—sometimes between the shoulders, sometimes in the side, or breast, etc. System extremely irritable, pulse hard and easily excited, from about ninety to one hundred and fifty, by the stimulus of a very small quantity of food; and, to be short, I was given up, on all hands, as lost.

From reading "Rush" I was induced to try a milk diet, and succeeded in regaining my health, so that for twenty-four years I have been entirely free from any symptom of phthisis; and although subject, during that time, to many attacks of fever and other epidemics, have steadily followed the business of a country physician.

I would further remark, before proceeding to the direct answer to your questions, that soon perceiving the benefit resulting from the course I had commenced, and finding the irritation to diminish in proportion as I diminished not only the quality, but quantity of my food, I took less than half a pint at a meal, with a small piece of bread, amounting to about the quantity of a Boston cracker; and at times, in order to lessen arterial action, added some water to the milk, taking only my usual quantity in *bulk*.

A seton was worn in the side, and a little exercise on horseback taken three times every day, as strength would allow, during the whole progress. The appetite was, at all times, not only *craving*, it was *voracious*; insomuch that all my sufferings from all other sources, dwindled to a point when compared with it.

The quantity that I ate at a time so far from satisfying my appetite, only served to increase it; and this inconvenience continued during the whole term, without the least abatement;—and the only means by which I could resist its cravings, was to live entirely by myself, and keep out of sight of all kinds of food except the scanty pittance on which I subsisted. And now to the proposed questions.

1. Increased.

2. More agreeable, hunger excepted.

3. To the first part of this question, I should say evidently clearer; to the latter part, such was the state of debility when I commenced, and such was it through the whole course, I am not able to give a decisive answer.

4. This question, you will perceive, is already answered in my preliminary remarks.

5. Fewer, insomuch that I had none.

6. Two full years.

7. My living, from early life, had been conformable to the habits of the farmers of New England, from which place I emigrated, and my habits in regard to stimulating drinks were always moderate; but I occasionally took them, in conformity to the customs of those "*times of ignorance.*"

8. I literally drank *nothing*; the milk wholly supplying the place of all liquids.

9. State of the bowels good before adopting the course, and after.

10. I do not.

11. I have not.

CALEB BANNISTER.

LETTER VII.—FROM DR. LYMAN TENNY.

FRANKLIN, Vermont, June 22, 1835.

SIR,—In answer to your inquiries, in the Boston Medical and Surgical Journal, vol. xii., page 78, I can say that I have lived entirely upon a bread and milk diet, without using any animal food other than the milk.

1. At first, my bodily strength was diminished to a certain degree, and required a greater quantity of food, and rather oftener, than when upon a mixed diet of animal food (strictly so called) and vegetables.

2. The animal sensations, attending upon the process of digestion, were rather more agreeable than when upon a mixed diet.

3. My mind was more clear, but I could not continue a laborious investigation as long as when I used animal food more plentifully.

4. At this time there were no constitutional infirmities which I was laboring under, except those which more or less accompany the rapid growth of the body; such as a general lassitude, impaired digestion, etc., which were neither removed nor aggravated, but kept about so, until I ate just what I pleased, without any regard to my indigestion, etc., when I began to improve in the strength of my whole system.

5. I do not recollect whether I was subject to more or fewer colds; but I can say I was perfectly free from all febrile attacks, although febrile diseases often prevailed in my vicinity. But since that time, a period of six years, I have had three attacks of fever.

6. The length of time I was upon this diet was about two years.

7. Before entering upon this diet, I was in the habit of taking a moderate quantity of animal food, but without very high seasoning or stimulants.

8. While using this diet, I confined myself entirely and exclusively to cold water as a drink—using neither tea, coffee, nor spirits of any kind whatever.

9. I am inclined to think that a vegetable diet is more aperient than an animal one; indeed, I may say I know it to be a fact.

10. From what I have experienced, I do not think that laborers would be any more healthy by excluding animal food from their diet entirely; but I believe it would be much getter if they would use less. As to students, I believe their health would be promoted if they were to exclude it almost, if not entirely.

11. I never have selected any vegetables which I thought to be more healthy than others: nor indeed do I believe there is any one that is more healthy than another; but believe that all those vegetables which we use in the season of them, are adapted to supply and satisfy the wants of the system.

We are carnivorous, as well as granivorous animals, having systems requiring animal, as well as vegetable food, to keep all the organs of the body in tune; and perhaps we need a greater variety than other animals.

Yours, etc.,

LYMAN TENNY.

LETTER VIII.—FROM DR. J. M. B. HARDEN.

LIBERTY COUNTY, Georgia, July 15, 1835.

SIR,—Having observed, in the May number of the "American Journal of the Medical Sciences," certain inquiries in relation to diet, proposed by you to the physicians of the United States, I herewith transmit to you an account of a case exactly in point, which I hope may prove interesting to yourself, and in some degree "assist in the settlement of a question of *great interest* to the *country*."

The case, to which allusion is made, occurred in the person of a very intelligent and truly scientific gentleman of this county, whose regular habits, both of mind and body, added to his sound and discriminating judgment, will tend to heighten the value and importance of the experiment involved in the case I am about to detail.

Before proceeding to give his answers to your interrogatories, it may be well to premise, that at the time of commencing the experiment, he was forty-five years of age; and being an extensive cotton planter, his business was such as to make it necessary for him to undergo a great deal of exercise, particularly on foot, having, as he himself declares, to walk seldom less than ten miles a day, and frequently more; and this exercise was continued during the whole period of the experiment. His health for two years previously had been very feeble, arising, as he supposed, from a diseased *spleen*; which organ is at this time enlarged, and somewhat indurated. His digestive powers have *always* been *good*, and he had been in the habit of making his meals at times entirely of *animal food*. His bowels have always been regular, and rather inclined to looseness, but never disordered. He is five feet eight inches high, of a very thin and spare habit of body, with thin dark hair, inclining to baldness; complexion rather dark than fair; eyes dark hazel; of *very studious* habits when free from active engagements; with great powers of mental abstraction and attention, and of a temper *remarkably even*.

In answer to your interrogatories, he replies,—

1. That his bodily strength was increased, and general health became better.

2. He perceived no difference.

3. He is assured of the affirmative.

4. His spleen was diminished in size, and frequent and long-continued attacks of *lumbago* were rendered *much milder*, and have so continued.

5. Had fewer colds and febrile attacks.

6. Three years.

7. No; with the slight exception mentioned above.

8. No.

9. In his case rather less.

10. Undoubtedly.

11. No; has made his meals of cabbages entirely, and found them as easily digested as any other article of diet. I may remark, that *honey* to him is a poison, producing, *invariably*, symptoms of cholera.

After three years' trial of this diet, without having any previous apparent disease, but on the contrary as strong as usual, he was taken, somewhat suddenly, in the winter of 1832 and 3, with symptoms of extreme debility, attended with œdematous swellings of the lower extremities, and painful cramps, at night confined to the gastrocnemii of both legs, and some feverishness, indicated more by the beatings of the *carotids* than by any other symptom. His countenance became very pallid, and indeed he had every appearance of a man in a very low state of health. Yet, during the whole period of this apparent state of disease, there were no symptoms indicative of disorder in any function, save the general function of innervation, and perhaps that of the lymphatics or absorbents of the lower extremities. Nor was there any manifest disease of any organ, unless it was the spleen, which was not then remarkably enlarged. I was myself disposed to attribute his symptoms to the spleen, and possibly to the want of animal food; but he himself attributes its commencement, if not its continuance, to the inhalation of the vapor of arseniuretted and sulphuretted hydrogen gases, to which he was subjected during some chemical experiments on the ores of cobalt, to which he has been for a long time turning his attention; a circumstance which I had not known until lately.

However it may be, he again returned to a mixed diet (to which however he ascribes no agency in his recovery), and, after six months' continuance in this state, he rapidly recovered his usual health and strength, which, up to this day—two full years after the expiration of six months—have continued good. In the treatment of his case no medicine of any kind was given, to which any good effect can be attributed; and indeed he may be said to have undergone no medical treatment at all.

Yours, etc.,

J. M. B. Harden.

LETTER IX.—FROM JOSEPH RICKETSON, ESQ.

New Bedford, 8th month, 26th, 1835.

Respected Friend,—Perhaps before giving answers to thy queries in the American Journal of Medical Science, it may not be amiss to give thee some account of my family and manner of living, to enable thee to judge of the effect of a vegetable diet on the constitution.

I have a wife, a mother aged eighty-eight, and two female domestics. It is now near three years since we adopted what is called the Graham or vegetable diet, though not in its fullest extent. We exclude animal food from our diet, but sometimes we indulge in shell and other fish. We use no kind of stimulating liquors, either as drink or in cookery, nor any other stimulants except occasionally a little spice. We do not, as Professor Hitchcock would recommend, nor as I believe would be most conducive to good health, live entirely simple; sometimes, however, for an experiment, I have eaten only rice and milk; at other times only potatoes and milk for my dinner; and have uniformly found I could endure as much fatigue, and walk as far without inconvenience, as when I have eaten a greater variety. We, however, endeavor to make our varieties mostly at different meals.

For breakfast and tea we have some hot water poured upon milk, to which we add a little sugar, and cold bread and butter; but in cold weather we toast the bread, and prefer having it so cool as not to melt the butter. We seldom eat a meal without some kind of dried or preserved fruit, such as peaches, plums, quinces, or apples; and in the season, when easily to be procured, we use, freely, baked apples, also berries, particularly blackberries stewed, which, while cooking, are sweetened and thickened a little. Our dinners are nearly the same as our other meals, except that we use cold milk, without any water. We have puddings sometimes made of stale bread, at others of Graham or other flour, or rice, or ground rice, usually baked; we have also hasty puddings, made of Indian meal, or Graham flour, which we eat with milk or melted sugar and cream; occasionally we have other simple puddings, such as tapioca, etc. Custards, with or without a crust, pies made of apple, and other fruits either green or preserved; but we have no more shortening in the crust than just to make it a little tender.

I have two sons; one lived with us about fifteen months after we adapted this mode of living; it agreed remarkably well with him; he grew strong and fleshy. He married since that time, and, in some measure, returned to the usual manner of living; but he is satisfied it does not agree so well with him as the Graham diet. The coarse bread he cannot well do without. My other son was absent when we commenced this way of living; he has been at home about six weeks, and has not eaten any animal food except when he dined out. He has evidently *lost* flesh, and is not very well; *he* thinks he shall not be able to live without animal food, but I think his indisposition is more owing to the season of the year than diet. He never drank any tea or coffee until about four years since, when he took some coffee for a while, but no tea. For the last two years he has not drank either, when he could get milk. He is generally healthy, and so is his brother: both were literally brought up on gingerbread and milk, never taking animal food of choice, until they were fifteen or sixteen years of age.

Dr. Keep, of Fairhaven, Connecticut, was here about a year since, in very bad health, since which I learn he has recovered by abstaining from animal food and other injurious diet. As he is a scientific man, I think he can give thee some useful information.

1. The strength of both myself and wife has very materially increased, so that we can now walk ten miles as easily as we could five before; possibly it may in part be attributed to practice. Our health is, in every respect, much improved. One of our women enjoys perfect health; the other was feeble when we commenced this way of living, and she has not gained much if any in the time; but this may be owing to her attendance on my mother, both day and night, who, being blind and feeble, takes no exercise except to walk across the room; but we are very sure she would not have lived to this time had she not adopted this way of living.

2. The process of digestion is much more agreeable, if we do not indulge in eating too much. We seldom have occasion to think of it after rising from the table.

3. I do not perceive much effect on the mind, other than what would naturally be produced by the restoration of health; but have no doubt a laborious investigation might be continued as long, if not longer, on this than any other diet.

4. I was formerly very much afflicted with the headache, and sometimes was troubled with rheumatism. I have very seldom, for the last two years especially, been troubled with either; and when I have had a turn of headache, it is light indeed compared with what it was before we adopted this system of living. My wife was very dyspeptic, and often had severe turns of palpitation of the heart; the latter is entirely removed, and she seldom experiences any inconvenience from the former. Our nurse was formerly, and still is, troubled with severe turns of headache, though not so bad as formerly; and I think she would have much less of it if she were placed in a different situation.

5. We scarcely know what it is to have a cold; my wife in particular. Previously to our change of diet, I was very subject to severe colds, attended with a hard cough, which lasted, sometimes, for several weeks.

6. As before stated, we exclude animal food from our diet, as well as tea and coffee.

7. Before we adopted a vegetable diet, we always had meat for dinner, and generally with breakfast; and not unfrequently with tea. Tea and coffee we drank very strong.

8. We have substituted milk and water sweetened, for tea and coffee.

9. Most vegetables I find have a tendency (especially when Graham or unbolted wheaten flour is used) to keep the bowels open; to counteract which, we use rice once or twice a week. Potatoes, when eaten freely, are flatulent, but not inconvenient when eaten moderately.

10. I think the health of students, by the exclusion of animal food from their diet, would be promoted, especially if they excluded tea and coffee also; and I can see no good reason why it should not be beneficial to laboring people. I have conversed with two or three mechanics, who confirm me in this belief.

11. Graham bread, as we call it, eaten with milk, or baked potatoes and milk, for most people, I think would be healthy; to which should be added such a proportion of rice as may be found necessary.

Thy friend,
JOSEPH RICKETSON.

LETTER X.—FROM JOSEPH CONGDON, ESQ.

NEW BEDFORD, Sept., 1835.

ANSWERS to Dr. North's inquiries on diet.

1. Increase of strength and activity, connected with, and perhaps in some good degree a consequence of, an increase of daily exercise.

2. Process of digestion more regular and agreeable.

3. Mental activity greater; no decisive experiments on the ability to *continue* a laborious investigation.

4. Dyspepsia of long continuance, and also difficult breathing; inflammation of the eyes.

5. Fewer colds; febrile attacks very slight; great elasticity in recovering from disease. Some part of the effect should undoubtedly be ascribed to greater attention to the skin by bathing and friction.

6. Twenty-six months of *entire abstinence* from all animal substances, excepting butter and milk. Salt is used regularly.

7. Through life inclined to a vegetable diet, with few stimulants.

8. Drinks have been milk, milk and water, or cold water.

9. A *well-selected* vegetable diet appears to produce a very regular action of the stomach and bowels.

10. I think the health of laborers and students would be promoted by a *great* reduction of the usual quantity of animal food, and perhaps by discontinuing its use entirely. I feel no want.

11. From my experience, I can very highly recommend bread made of coarse wheat flour. Among fruits, the blackberry, as peculiarly adapted to the state of the body, at the time of the year when it is in season. My range of food has been confined. I avoid green vegetables. Age 35.

JOSEPH CONGDON.

LETTER XI.—FROM GEORGE W. BAKER, ESQ.

NEW BEDFORD, 9th month, 10, 1835.

DR. M. L. NORTH,—Agreeably to request, the following answers are forwarded, which I believe to be correct as far as my experience has tested.

1. At first it was diminished; but after a few months it was restored, and I think increased.

2. More.

3. It could.

4. Pretty free from constitutional infirmities before the change, and no increase since.

5. I have had no cold, of any consequence, for the last three years; at which time I substituted cold water for tea and coffee, and commenced using cold water for washing about my head and neck and for shaving, which I continued through the year.

6. I have not eaten animal food for about eighteen months.

7. Two years previous to the entire change the quantity was great, but there had been a gradual diminution.

8. It was. (See fifth answer.)

9. More so, in my case.

10. I believe the health of both laborers and students would be improved.

11. I have generally avoided eating cucumbers; otherwise I have not.

Thy assured friend,

GEO. W. BAKER.

LETTER XII—FROM JOHN HOWLAND, JR., ESQ.

NEW BEFORD, 9th month, 10th day, 1835.

FRIEND,—As I have lived nearly three years upon a vegetable diet, I cheerfully comply with thy request.

1. My bodily strength has been increased; and I can now endure much more exercise than formerly, without fatigue.

2. They are more agreeable; and I am now free from that dull, heavy feeling, which I used to experience after my meals.

3. My mind is much clearer; and I am free from that depression of spirits, to which I was formerly subject.

4. I was of a costive, dyspeptic habit, which has been entirely removed. I had frequent and severe attacks of headache, which I now rarely have; and when they do occur they are very light, compared with what they formerly were.

5. I have had fewer colds, and those much lighter than formerly.

6. About three years.

7. I used to eat animal food for breakfast and dinner, with coffee for drink, at those meals; and tea for my third meal, with bread and butter.

8. Milk for breakfast, and cold water for the other two meals.

9. I have found it more so; inasmuch as the use of it, with the substitution of bread, made from *coarse, unbolted wheat flour*, instead of superfine, has removed my costiveness entirely.

10. I do.

11. I consider potatoes and rice as the most healthy, and confine myself principally to the former.

I would remark that during the season of fruits, I eat freely of them, with milk; and consider them to be healthy.

JOHN HOWLAND, JR.

LETTER XIII.—FROM DR. W. H. WEBSTER.

BATAVIA, N. Y., Oct. 21, 1835.

SIR,—Some months since, I read your inquiries on diet in the Boston Medical and Surgical Journal; and subsequently in the Journal of Medical Sciences, Philadelphia.

I will answer your questions, numerically, from my knowledge of a case somewhat in point, and with which I am but too familiar, as it is my own. But, first, let me premise a few points in the history of my health, as a kind of key to my answers.

It is about fifteen years since I was called a *dyspeptic*; this was while engaged in my academical studies. Not being instructed by my medical friend to make any alteration in diet and regimen, I merely swallowed his cathartics for one month, and his anodynes for the next month, as the bowels were constipated or relaxed. In short, I left college more dead than alive—a confirmed dyspeptic.

In 1826, I commenced the practice of physic. From this time, to the winter of 1831-2, I found it necessary gradually to diminish my indulgence in the luxuries of the table—especially in animal food, and distilled and fermented liquors. On one of the most inclement nights of the winter of 1831-2, a fire broke out in our village, at which I became very wet by perspiration, and the ill-directed efforts of some to extinguish it. This was followed by a severe inflammatory attack upon the digestive organs generally, and especially upon the renal region, which confined me to the house for more than eight months; and, for the greatest share of that time, with the most excruciating torture. On getting out again, I found myself in a wretched condition indeed—reduced to a skeleton—a voracious appetite, which could not be indulged, and which had scarcely deserted me through the whole eight months. I could not regain my flesh or strength but by almost imperceptible degrees; indeed, loaf-sugar and crackers were almost the only food I could use with impunity for the first year.

It is now nearly four years since I have eaten animal food, unless it be here and there a little, as an experiment, with the sole exception of oysters, in which I can indulge, but with all due deference to the stricter rules of temperance. Still my appetite for animal food seems unabated. I have ever been a man unusually temperate in the use of intoxicating drinks; and by no means intemperate in the luxuries of the table. I take no meat, no alcoholic or fermented drinks, not even cider; and, for a year past, my health has been better than for three years previous; and I think that about one third the amount of nourishment usually taken by men of my age, might subserve the purposes of food for *me* better than a larger quantity. The more I eat, the more I desire to eat; and abstinence is my best medicine.

But I have already surpassed my limits, and here are my answers.

1. My strength is invariably diminished by animal food, and in almost direct proportion to the quantity, with the exception named above.

2. Pain has been the uniform attendant upon the digestion of an animal diet, with feverish restlessness and constipation.

3. Decidedly more fit for energetic action.

4. An irritation, or subacute inflammation of the digestive apparatus, which is aggravated by animal food.

5. Can endure hardship, exposure, and fatigue, much better without meat.

6. About four years, with the exception stated above.

7. It was not.

8. Partially at the commencement; but not of late, if not taken hot.

9. Much more aperient.

10. Both classes take too much; and students and sedentaries should take little or none.

11. For myself farinaceous articles first, then the succulent sub-acid ripe fruits, then the less oily nuts are most healthful—and animal food, strong coffee and tea, and unripe or hard fruits, in any considerable quantities, are most pernicious.

Yours, etc.,

W. H. WEBSTER.

LETTER XIV.—FROM JOSIAH BENNET, ESQ.

MOUNT-JOY, Pa., Oct. 27, 1835.

SIR,—I hereby transmit to you, answers to a series of dietetic queries which you have recently submitted.

1. My physical strength was at least equal (I am rather inclined to think greater) after abstaining from animal food. I was, I am certain, not subject to such general debility and lassitude of the system, after considerable bodily exercise.

2. More agreeable—not being subject to a sense of vertigo, which frequently (with me) followed the use of animal food. There is, generally, more cheerfulness and vivacity.

3. The mind is more clear, and is not so liable to be confused when intent upon any intricate subject; and, of course, "can continue a laborious investigation longer." There is at no time such a propensity to incogitancy.

4. I am not aware of being the subject of any "constitutional infirmities;" yet, that the change of diet had a very great effect upon the system, is obvious, from the fact of my having been, formerly, subject to an eruptive disease of the skin, principally on the shoulders and upper part of the back, for a number of years, which is not the case at present, nor do I think will be, as long as I continue my present mode of living.

5. I think I have not had as many colds and febrile attacks as before, nor have they been so severe; yet I cannot be very decisive on this point, on account of the length of time in the trial not being fully sufficient.

6. Between seven and eight months. I must here state that animal food was not *entirely* excluded. I probably partook, in very moderate quantities, once or twice a week.

7. The quantity of animal food which would be considered "an uncommon proportion," I am unable to determine; but I was accustomed to make use of it, not *less* than twice, and sometimes three times a day, moderately seasoned. No other stimulants, of any account.

8. Cold water has been the only substitute for tea and coffee, with the exception of an occasional cup; probably as often as once or twice a week. I was, on several occasions, by personal experience, induced to believe that the use of strong coffee retarded the process of digestion.

9. More aperient. Previous to the general exclusion of animal food from my diet, I was subject to inveterate costiveness; cases of which are now neither frequent nor severe.

10. I do firmly believe it would.

11. My diet, principally, during the trial, consisted of wheat bread, of the proper age, with a moderate quantity of fresh butter. Potatoes, beans, and some other esculent roots, etc., I found to be nutritious and healthy. The following substances I found to produce a contrary effect, or to possess different qualities: cabbage, when not well boiled; cucumbers, raw or pickled; radishes, beets, and the whole catalogue of preserves. Fresh bread was particularly hurtful to me.

Yours, etc.,

JOSIAH BENNETT.

LETTER XV.—FROM WILLIAM VINCENT, ESQ.[2]

HOPKINTON, R. I., Dec. 23, 1835.

SIR,—The following answer to the interrogations in the Boston Medical and Surgical Journal of March 1835, on diet, etc., as proposed by yourself, has been through the press of business, neglected until this late period. Trusting they may be of some use, I now forward them.

1. Rather increased, if any change.

2. ——

3. I think I have retained the vigor of my mind more, in consequence of an abstemious diet.

4. I thought I had the appearance of scurvy, which gradually disappeared.

5. ——

6. From May 20, 1811, (more than twenty-four years.)

7. Small in quantity, and dressed and cooked simply.

8. I have drank nothing but warm tea, for seven years.

9. Bowels uniformly open.

10. I should not think it would.

11. I have lived principally on bread, butter, and cheese, and a few dried vegetables.

I was born March 31, 1764. In 1833, when mowing, to quench thirst, I drank about a gill of cold water, *after* about as much milk and water; and the same year, some molasses and water; but they did not answer the purpose. But when I rinsed my mouth with cold water, it allayed my thirst.

(Signed)

WM. VINCENT.

LETTER XVI.—FROM L. R. BRADLEY, BY DR. GEO. H. PERRY.

HOPKINTON, R. I., Dec. 23, 1835.

SIR,—I deem it necessary, first, to mention the situation of my health, at the time of commencing abstinence from animal food. I was recovering from an illness of a *nervous fever*. A sudden change respecting my food not sitting well, rendered it necessary for me to abstain from all kinds, excepting dry wheat bread and gruel, for several weeks. By degrees I returned to my former course of diet, but as yet not to its full extent, as I cannot partake of animal food of any kind whatever, nor of vegetables cooked therewith.

1. Diminished.

2. ——

3. I do not perceive the mind to be clearer, and the power of investigation less.

4. Distress in the stomach and pain in the head removed.

5. ——

6. Six years and ten months.

7. Unusual proportion of animal food.

8. The first year, I drank only warm water, sweetened; since that, tea.

9. ——

10. I do not.

11. I find *beets* particularly hard to digest.

L. R. B.

The foregoing statements and answers are in her own way and manner.

Yours, etc.,

Geo. H. Perry.

LETTER XVII.—FROM DR. L. W. SHERMAN.

Falmouth, Mass., March 28, 1835.

Sir,—In compliance with the request you recently made in the Medical Journal, I inclose the following answers to the queries relative to regimen you have propounded. They are given by a lady, whose experience, intelligence, and discernment, have eminently qualified her to answer them. She, with myself, is equally interested with you in having this important question settled, and is extremely happy that you have undertaken to do it. This lady is now fifty years of age; her constitution naturally is good; her early habits were active, and her diet simple, until twenty years of age. After that, until within a few years, her living consisted of all kinds of meats and delicacies, with wine after dinners, etc., etc.

1. Her bodily strength was greatly increased by excluding animal food from her diet.

2. The animal sensations connected with the process of digestion have been decidedly more agreeable.

3. The mind is much clearer, the spirits much better, the temper more even, and "less irritability pervades the system." The mind can continue a laborious investigation longer than when she subsisted on a mixed diet.

4. Her health, which was before feeble, has, by the change, been decidedly improved.

5. She has certainly had fewer colds, and no febrile attacks of any consequence, since she has practiced rigid abstinence from meats.

6. She has abstained entirely for three years, and has taken but little for seven or eight years; and whenever she has, from necessity (in being from home, where she could procure nothing else), indulged in eating meat, she has universally suffered severely in consequence.

7. The change to a vegetable diet was preceded, in her case, by the use of an uncommon proportion of animal food, highly seasoned with stimulants.

8. Tea and coffee she has not used for thirteen years. She has used, for substitutes, water, milk and water, barley water, and gruel. She found tea and coffee to have an exceedingly pernicious effect upon her nervous and digestive system.

9. A vegetable diet is more aperient than a mixed. Habitual constipation has been entirely removed by the change.

10. She sincerely believes, from her experience, that the health of laborers and students would be generally promoted by the exclusion of animal food from their diet.

11. She considers *hominy*, as prepared at the South, particularly healthy; and subsists upon this, with bread made from coarse flour, with broccoli, cauliflower, and all kinds of vegetables in their season.

Be assured, dear sir, that these answers have come from a high source, to which private reference may at any time be made, and consequently are entitled to the highest consideration.

Yours, etc.,

L. W. SHERMAN.

NOTE.—If I have not been minute enough in the relation of this case, I shall hereafter be happy to answer any questions you may think proper to propose. It is a very interesting and important case, in my opinion. The lady has been under my care a number of times, while laboring under slight indisposition. She has always been very regular and systematic in all her habits. She is healthy and robust in appearance, and looks as though she might not be more than forty. This is the only case of the kind within my knowledge. I have practiced on her plan for a few weeks at a time, and, so far as my experience goes, it precisely comports with hers. But I love the "good things" of this world too well to abstain from their use, until some formidable disease demands their prohibition.

Yours, etc.,

L. W. S.

FOOTNOTES:

[1] Dr. Preston has since deceased.

[2] Mr. Vincent is of Stonington, Ct.

CHAPTER III.

REMARKS ON THE FOREGOING LETTERS.

Correspondence.—The "prescribed course of Regimen."—How many victims to it?—Not one.—Case of Dr. Harden considered.—Case of Dr. Preston.—Views of Drs. Clark, Cheyne, and Lambe, on the treatment of Scrofula.—No reports of Injury from the prescribed System.—Case of Dr. Bannister.—Singular testimony of Dr. Wright.—Vegetable food for Laborers.—Testimony, on the whole, much more favorable to the Vegetable System than could reasonably have been expected, in the circumstances.

"Reports not unfrequently reach us," says Dr. North, "of certain individuals who have fallen victims to a prescribed course of regimen. These persons are said, by gentlemen who are entitled to the fullest confidence, to have pertinaciously followed the course, till they reached a point of reduction from which there was no recovery." "If these are facts," he adds, "they ought to be known and published."

It was in this view, that Dr. North, himself a medical practitioner of high respectability, sent forth to every corner of the land, through standard and orthodox medical journals, to regular and experienced physicians—his "medical brethren"—his list of inquiries. These inquiries, designed to elicit truth, were couched in just such language as was calculated to give free scope and an acceptable channel for the communication of every fact which seemed to be opposed to the VEGETABLE SYSTEM; for this, we believe, was distinctly understood, by every medical man, to be the "prescribed course of regimen" alluded to.

The results of Dr. North's inquiries, and of an opportunity so favorable for "putting down," by the exhibition of sober facts, the vegetable system, are fully presented in the foregoing chapter. Let it not be said by any, that the attempt was a partial or unfair one. Let it be remembered that every effort was made to obtain *truth in facts*, without partiality, favor, or affection. Let it be remembered, too, that nearly two years elapsed before Dr. North gave up his papers to the author; during which time, and indeed up to the present hour—a period, in the whole, of more than fourteen years—a door has been opened to every individual who had any thing to say, bearing upon the subject.

Let us now review the contents of the foregoing chapter. Let us see, in the first place, what number of persons have here been reported, by medical men, as having fallen victims to the said "prescribed course of regimen."

The matter is soon disposed of. Not a case of the description is found in the whole catalogue of returns to Dr. N. This is a triumph which the friends of the vegetable system did not expect. From the medical profession of this country, hostile as many of them are known to be to the "prescribed course of regimen," they must naturally have expected to hear of at least a few persons who were supposed to have fallen victims to it. But, I say again, not one appears.

It is true that Dr. Preston, of Plymouth, Mass., thinks he should have fallen a victim to his abstinence from flesh meat, had he not altered his course; and Dr. Harden, of Georgia, relates a case of sudden loss of strength, and great debility, which he thought, *at the time*, might "possibly" be ascribed to the want of animal food: though the individual himself attributed it to quite another cause. These are the only two, of a list of thirty or forty, which were detailed, that bear the slightest resemblance to those which report had brought to the ear of Dr. N., and about which he so anxiously and earnestly solicited inquiry of his medical brethren.

As to the case mentioned by Dr. Harden, no one who examined it with care, will believe for a moment, that it affords the slightest evidence against a diet exclusively vegetable. The gentleman who made the experiment had pursued it faithfully three years, without the slightest loss of strength, but with many advantages, when, of a sudden, extreme debility came on. Is it likely that a diet on which he had so long been doing well, should produce such a sudden falling off? The gentleman himself appears not to have had the slightest suspicion that the debility had any connection with the diet. He attributes its commencement, if not its continuance, to the inhalation of poisonous gases, to which he was subjected in the process of some chemical experiments.

But why, then, it may be asked, did he return to a mixed diet, if he had imbibed no doubts in regard to a diet exclusively vegetable; and, above all, how happened he to recover on it? To this it may be replied, that there is every reason to believe, from the tenor of the letter, that he acted against his own inclination, and contrary to his own views, at the request of his friends, and of Dr. Harden, his physician; though Dr. Harden does not expressly say so. Besides, it does not appear that under his mixed diet there was any favorable change, till something like six months had elapsed. This was a period, in all probability, just sufficient to allow the poison of the gases to disappear; after which he might have been expected to recover on any diet not positively bad. If this is not a true solution of the case, how happens it that there was no disease of any organ or function, except the nervous function? There is every reason for believing that Dr. Harden, at the date of his letter, had undergone a change of opinion, and was himself beginning to doubt whether the regimen had any agency in producing the debility.[3]

The case of Dr. Preston is somewhat more difficult. At first view, it seems to sustain the old notion of medical men, that, with a scrofulous habit, a diet exclusively vegetable cannot be made to agree. This, I say, seems to be a natural conclusion, *at first view*. But, on looking a little farther, we may find some facts that justify a different opinion.

Dr. Preston was evidently timid and fearful—foreboding ill—during the whole progress of his experiment. We think his story fully justifies this conclusion. In such circumstances, what could have been expected? There is no course of regimen in the world which will succeed happily in a state of mind like this.

It should be carefully observed by the reader, that Dr. Preston speaks of entering upon a "severe course of diet;" and also, that, in attempting to give an opinion as to the best kind of vegetable food, he speaks of potatoes, prepared in a certain specified manner, as being preferable to any other. Now, I think it obvious, that Dr. Preston's "severe course" partook largely of *crude* vegetables, instead of the richer and better farinaceous articles—as the various sorts of bread, rice, pulse, etc.—and, if so, it is not to be wondered at that it was so unsuccessful. In short, I do not think he made any thing like a fair experiment in vegetable diet. His testimony, therefore, though interesting, seems to be entitled to very little weight.

This conclusion is stated with the more confidence, from the fact that some of the best medical writers, not only of ancient times, but of the present day, appear to entertain serious doubts in regard to the soundness of the popular opinion in favor of the "beef-steak-and-porter" system of curing scrofulous patients. Dr. Clark, in the progress of his "Treatise on Consumption," almost expresses a belief that a judicious vegetable diet is preferable even for the scrofulous. He would not, of course, recommend a diet of *crude* vegetables, but one, rather, which would partake largely of farinaceous grains and fruits. Nor do I suppose he would, in every case, entirely exclude milk.

Dr. Cheyne, in his writings, not only gives it as his opinion that a milk diet, long continued, or a milk and vegetable diet and mild mercurials, are the best means of curing scrofula; but he also says, expressly, that "in all countries where animal food and strong fermented liquors are too freely used, there is scarcely an individual that hath not scrofulous glands." A sad story to relate, or to read! But, Dr. Lambe, of London, and other British physicians, entertain similar sentiments; and Dr. Lambe practices medicine largely, while entertaining these sentiments. I could mention more than one distinguished physician, in Boston and elsewhere, who prescribes a vegetable and milk diet in scrofula.

But, granting even the most that the friends of animal food can claim, what would the case of Dr. Preston prove? That the healthy are ever injured by the vegetable system? By no means. That the sickly would generally be? Certainly not. Dr. Preston himself even specifies one disease, in which he thinks a vegetable diet would be useful. What, then, is the bearing of *this single and singular case?* Why, at the most, it only shows that there are some forms of dyspepsia which require animal food. Dr. Preston does not produce a single fact unfavorable to a diet exclusively vegetable for the healthy.[4]

It is also worthy of particular notice, that not a fact is brought, or an experiment related, in a list of from thirty to forty cases, reported too by medical men, which goes to prove that any injury has arisen to the healthy, from laying aside the use of animal food. This kind of information, though not the principal thing, was at least a secondary object with Dr. North; as we see by his questions, which were intended to be put to those who had excluded animal food from their diet for a year or more.

But, let us take a general view of the replies to the inquiries of Dr. North. The sum of his first three questions, was,—What were the effects of excluding animal food from your diet on your bodily strength, your mental faculties, and your appetite and animal spirits?

The answers to the three questions, of which this is the same, are, as will be seen, remarkable. In almost every instance the reply indicates that bodily and mental labor was endured with less fatigue than before, and that an increased activity of mind and body was accompanied with increased cheerfulness and animal enjoyment. In nearly every instance, strength of body was actually increased; especially after the first month. A result so uniformly in favor of the vegetable system is certainly more than could have been expected.

One physician who made the experiment, indeed, says, that though his mind was clearer than before, he could not endure, so long, a laborious investigation. Another individual says, he perceived no difference in this respect. A third says, she found her bodily strength and powers of investigation somewhat diminished, though her disease was removed. With these exceptions, the testimony on this point is, as I have already said, most decidedly—I might say most overwhelmingly—in favor of the disuse of animal food.

To the question, whether any constitutional infirmities were aggravated or removed by the new course of regimen, the replies are almost equally favorable to the vegetable system. It is true that one of the physicians, Dr. Parmly, thinks the beneficial effects which appeared in the circle of his observation were the results of a simultaneous discontinuance of fermented drinks, tea and coffee, and condiments. But I believe every one who reads his letter will be surprised at his conclusions. No matter, however; we have his facts, and we are quite willing they should be carefully considered. The singular case of Dr. Preston, I now leave wholly out of the account. It was, as I have since learned, the story of a *very singular man.*

Among the diseases and difficulties which were removed, or supposed to be removed, by the new diet, were dyspepsia, with the constipation which usually attends it, general lassitude, rheumatism, periodical headache, palpitations, irritation of the first passages, eruptive diseases of the skin, scurvy, and consumption.

The case of Dr. Bannister, who was, in early life, decidedly consumptive, is one of the most remarkable on record. Though evidently consumptive, and near the borders of the grave, between the ages of twenty and twenty-nine, he so far recovered as to be, at the age of fifty-three, entirely free from every symptom of phthisis for twenty-four years; during which whole period, he was sufficiently vigorous to follow the laborious business of a country physician.

The confidence of Dr. Wright in the prophylactic powers of a diet exclusively vegetable, so far as the mere opinion of one medical man is to be received as testimony in the case, is also remarkable. He not only regards the vegetable system as a defence against the diseases of miasmatic regions, but also against the varioloid disease. On the latter point, he goes, it seems, almost as far as Mr. Graham, who appears to regard it not only as, in some measure, a preventive of epidemic diseases generally, in which he is most undoubtedly correct, but also of the small-pox.

The testimony on another point which is presented in the replies to Dr. North's questions, is almost equally uniform. In nearly every instance, the individuals who have abandoned animal food have found themselves less subject to colds than before; and some appear to have fallen into the habit of escaping them altogether. When it is considered how serious are the consequences of taking cold—when it is remembered that something like one half of the diseases of our climate have their origin in this source—it is certainly no trifling evidence in favor of a course of regimen, that, besides being highly favorable in every other respect, it should prove the means of freeing mankind from exposure to a malady at once troublesome in itself and disastrous in its consequences.

In reply to the question,—Is a vegetable diet more or less aperient than a mixed one,—the answers have been the same, in nearly every instance, that it is more so.

The answers to the question whether it was believed the health of either laborers or students would be promoted by the exclusion of animal food from their diet, are rather various. It will be observed, however, that many of the replies, in this case, are medical *opinions*, and come from men who, though they felt themselves bound to state facts, were doubtless, with very few exceptions, prejudiced against an exclusively vegetable regimen for the healthy. It is, therefore, to me, a matter of surprise, to find some of them in favor of the said prescribed course of regimen, both for students and laborers, and many of them in favor of the discontinuance of animal food by students. Those who have themselves made the experiment, with hardly an exception, are decidedly in favor of a vegetable regimen for all classes of mankind, particularly the sedentary. And in regard to the necessity of diminishing the proportion of animal food consumed by all classes, there seems to be but one voice.

On one more important point there is a very general concurrence of opinion. I allude to the choice of articles from the vegetable kingdom. The farinacea are considered as the best; especially wheat, ground without bolting. The preference of Dr. Preston is an exception; and there are one or two others.

On the whole—I repeat it—the testimony is far more favorable to the "prescribed course of regimen," both for the healthy and diseased than under the circumstances connected with the inquiry the most thorough-going vegetable eater could possibly have anticipated. If this is a fair specimen—and I know no reason why it may not be regarded as such—of the results of similar experiments and similar observations among medical men throughout our country, could their observations and experiments be collected, it certainly confirms the views which some among us have long entertained on this subject, and which will be still more strongly confirmed by evidence which will be produced in the following chapters. Had similar efforts been made forty or fifty years ago, to ascertain the views of physicians and others respecting the benefits or safety of excluding wine and other fermented drinks in the treatment of several diseases, in which not one in ten of our modern practitioners would now venture to use them, as well as among the healthy, I believe the results would have been of a very different character. The opinions, at least, of the physicians themselves, would most certainly have been, nearly without a dissenting voice, that the entire rejection of wine and fermented liquors was dangerous to the sick, and unsafe to many of the healthy, especially the hard laborer. And there is quite as much reason to believe that animal food will be discarded from our tables in the progress of a century to come, as there was, in 1800, for believing that all drinks but water would be laid aside in the progress of the century which is now passing.

FOOTNOTES:

[3] See a more recent letter from Dr. Harden, in the next chapter.

[4] Besides, it is worthy of notice, that Dr. Preston did not long survive on his own plan. He died about the year 1840.

CHAPTER IV.

ADDITIONAL INTELLIGENCE.

Letter from Dr. H. A. Barrows.—Dr. J. M. B. Harden.—Dr. J. Porter.—Dr. N. J. Knight.—Dr. Lester Keep.—Second letter from Dr. Keep.—Dr. Henry H. Brown.—Dr. Franklin Knox.—From a Physician.—Additional statements by the Author.

During the years 1837 and 1838 I wrote to several of the physicians whose names, experiments, and facts appear in Chapter II. Their answers, so far as received, are now to be presented.

I have also received interesting letters from several other physicians in New England and elsewhere—but particularly in New England—on the same general subject, which, with an additional statement of my own case, I have added to the foregoing. I might have added a hundred authentic cases, of similar import. I might also have obtained an additional amount of the same sort of intelligence, had it not been for the want of time, amid numerous other pressing avocations, for correspondence of this kind. Besides, if what I have obtained is not satisfactory, I have many doubts whether more would be so.

The first letter I shall insert is from Dr. H. A. Barrows, of Phillips, in Maine. It is dated October 10, 1837, and may be considered as a sequel to that written by him to Dr. North, though it is addressed to the author of this volume.

LETTER I.—FROM DR. H. A. BARROWS.

DEAR SIR,—As to food, my course of living has been quite uniform for the last two or three years—principally as follows. Wheat meal bread, potatoes, butter, and baked sweet apples for breakfast and dinners; for suppers, old dry flour bread, which, eaten very leisurely without butter, sauce, or drink, sits the lightest and best of any thing I eat. But I cannot make this my principal diet, because the bowels will not act (*without physic*) unless they have the spur of wheat bran two thirds of the time. I have at times practiced going to bed without any third meal; and have found myself amply rewarded for this kind of fasting, and the consequent respite thereby afforded the stomach, in quiet sleep and improved condition the next day. And as to drink, I still use cold water, which I take with as great a zest, and as keen a relish, as the inebriate does his stimulus. I seldom drink any thing with my meals; and if I could live without drinking any thing between meals, I think I should be rid of the principal "thorn in my side," the acetous fermentation so constantly going on in my epigastric storehouse.

As to exercise, I take abundance; perform all my practice (except in the winter) on horseback, and find this the very best kind of exercise for me. I seldom eat oftener than at intervals of six hours, and am apt to eat too much—have at various times attempted Don Cornaro's method of weighing food, but have found it rather dry business, probably on account of its conflicting with my appetite; but I actually find that my stomach does not bear watching at all well.

My brother continues to practice nearly total abstinence from animal food. I have seen him but once in two and a half years, but learn his health has greatly improved, so that he was able to take charge of a high school in the fall of 1836, of an academy in the spring of the present year, and also again this fall. During his vacation last July, he took a tour into the interior of Worcester county, Mass., and came home entirely on foot by way of the Notch of the White Hills, traveling nearly three hundred miles. This speaks something in favor of rigid abstinence—as when he commenced this regimen he was extremely low.

Yours sincerely,

H. A. BARROWS.

LETTER II.—FROM DR. JOHN M. B. HARDEN.

GEORGIA, Liberty Co., Oct. 19, 1837.

DEAR SIR,—I stated in my letter to Dr. North, if I recollect correctly, that the use of animal food was resumed in consequence of a protracted indisposition brought on, *as was supposed*, by the inhalation of arseniuretted hydrogen gas. The gentleman had begun to recover some time previously; and in a short time after he commenced the use of the animal food, he was restored to his usual health. He has continued the use of it ever since to the same extent as in the former part of his life. He has lately passed his fifty-fifth year, and is now in the enjoyment of as good health as he has ever known.

I know of a gentleman in an adjoining county, who with his lady has been living for some time past on a purely vegetable diet. They have not continued it long enough, however, to make the experiment a fair one.

No case of injury from the inhalation of arseniuretted hydrogen has come under my own personal observation, if we except the one above alluded to. I find, however, that Gehlen, a celebrated French chemist, fell a victim to it in the year 1815. His death is thus announced in the "Philosophical Magazine" for that year. "We lament to have to announce the death of Gehlen, many years the editor of an excellent Journal on Chemistry and other sciences, and a profound chemist. He fell a victim to his ardent desire to promote the advancement of chemical knowledge. He was preparing, in company with Mr. Rehland, his colleague, some arsenated hydrogen gas, and while watching for the full development of this air from its acid solution, trying every moment to judge from its particular smell when that operation would be completed, he inhaled the fatal poison which has robbed science of his valuable services." Vide Tillock's Phil. Mag., vol. 46, p. 316. Some further notice is taken of his death in a paper extracted from the "Annales de Chimie et de Physique," and published in a subsequent volume of the same Magazine. Vide vol. 49, p. 280, in which are given his last experiments on that subject, by M. Gay Lussac. I regret that no account is given in the same work of the symptoms arising from the poison in his case. I presume, however, they are on record.

In the subject of the case I mention, the general and prominent symptoms were an immediate and great diminution of muscular strength, with pallor of countenance and constant febricula, the arteries of the head beating with violence, particularly when lying down at night, the pulse always moderately increased in frequency, and full, but not tense; and digestion for the most part good. This state continued for about three months, during which time he was attending to his usual business, although not able to take as much exercise as before. At the end of this time he began to recover slowly, but it was six months before he was restored entirely.

Yours, etc.,

JOHN M. B. HARDEN.

LETTER III.—FROM DR. JOSHUA PORTER.

NORTH BROOKFIELD, Oct. 26, 1827.

Though I would by no means favor the propensity for book-making, so prevalent in our day, yet I have been long of the opinion that a work on vegetable diet for general readers was greatly needed. I need it in my family; and there are many others in this vicinity who would be materially benefited by such a work.

I have had no means of ascertaining the good or bad effects of a "diet exclusively vegetable in cases of phthisis, scrofula, and dyspepsia," for I have had none of the above diseases to contend with. But, since your letter was received, I have been called to prescribe for a man who has been a flesh eater for more than half a century. He was confined to his house, had been losing strength for several months, still keeping up his old habits. The disease which was preying upon him was chronic inflammation of the right leg; the flesh had been so long swollen and inflamed that it had become hard to the touch. There were ulcers on his thigh, and some had made their appearance on the hip. This disease had been of *seven months'* standing, though not in so aggravated a form as it now appeared. During this time, all the local applications had been made that could be thought of by the good ladies in the neighborhood; and after every thing of the kind had failed, they concluded to send for "the doctor."

After examining the patient attentively, I became convinced that the disease, which developed itself locally, was of a constitutional origin, and of course could not be cured by local remedies. All local applications were discontinued; the patient was put on a vegetable diet after the alimentary canal was freely evacuated. I saw this man three days afterward. The dark purple appearance of the leg had somewhat subsided; the red and angry appearance about the base of the ulcers was gone, his strength improved, etc. Three days after I called, I found him in his garden at work.

He is now—two weeks since my first prescription—almost well. All the ulcers have healed, with the exception of one or two. This man, who thinks it wicked not to use the good things God has given us—such as meat, cider, tobacco, etc.—is very willing to subsist, for the present, on vegetable food, because he finds it the only remedy for his disease.

Early in the spring of 1830, while a student at Amherst College, I was attacked with dyspepsia, which rendered my life wretched for more than a year, and finally drove me from college; but it had now so completely gained the mastery, that no means I resorted to for relief afforded even a palliation of my sufferings. After I had suffered nearly two years in this way, I was made more wretched, if possible, by frequent attacks of colic, with pains and cramps extending to my back; and so severe had these pains become, that the prescriptions of the most eminent physicians afforded only partial relief.

On the 13th of February, 1833, after suffering from the most violent paroxysm I had ever endured, I left my home for Brunswick, Maine, to attend a course of medical lectures. For several days I boarded at a public house, and ate freely of several substantial dishes that were before me. The consequence was a fresh attack of colic. From some circumstances that came up at this time, I was convinced that flesh meats had much to do with my sufferings, and the resolution was formed at once to change my diet and "starve" out dyspepsia.

I took a room by myself, and made arrangements for receiving a pint of milk per day; this, with coarse rye and Indian bread, constituted my only food. After living in this way a week or two, I had a free and natural evacuation. Thus nature began to effect what medicine alone had done for nearly three years. The skin became moist, and my voracious appetite began to subside. I returned home to my friends at the close of the term well, and have been well ever since—have never had a colic pain or any costiveness since that time. My powers of digestion are good, and though I do not live so rigidly now as when at Brunswick, I always feel best when my food is vegetables and milk. I can endure fatigue and exposure as well as any man. On this mild diet, too, my muscular strength has considerably increased; and every day is adding new vigor to my constitution.

Having experienced so much benefit from a mild diet, and being rationally convinced that man was a fruit-eating animal naturally, I made my views public by a course of lectures on physiology, which I delivered in the Lyceum soon after I came to this place (three years ago). The consequence was, that quite a number of those who heard my lectures commenced training their families as well as themselves to the use of vegetables, etc., and I am happy to inform you that, at this day, many of our most active influential business-doing men are living in the plainest and most simple manner.

One of my neighbors has taken no flesh for more than three years. He is of the ordinary height, and sanguine temperament, and usually weighed, when he ate flesh, one hundred and eighty pounds. After he changed his diet, his countenance began to change, and his cheeks fell in; and his meat-eating friends had serious apprehensions that he would survive but a short time, unless he returned to his former habits. But he persevered, and is now more vigorous and more athletic than any man in the region, or than he himself has ever been before.

His muscular strength is very great. A few days since, a number of the most athletic young men in our village were trying their strength at lifting a cask of lime, weighing five hundred pounds. All failed to do it, with the exception of one, who partly raised it from the ground. After they were gone, this vegetable eater without any difficulty raised the cask four or five times. More than three years ago this man lost his daughter, who fell a prey to cholera infantum; he has now a daughter rather more than a year old, whom he has trained on strictly physiological principles; and though very feeble at birth, and for three months subsequently, she is now the most healthy child in the town. This child had some of the first symptoms of consumption last August, owing to the too free indulgence of the mother in improper articles of food; but being treated with demulcents, at the same time correcting the mother's system, she recovered, and is now the "picture of health."

I was conversing with this gentleman the other day respecting his health—says he is perfectly well, weighs one hundred and sixty-five pounds; and though he was called well when eating flesh, he was not so in reality; for every few weeks he was troubled with headache and a sense of fullness in the region of the stomach, for which he was obliged to take an active cathartic. For a few months before he adopted the vegetable system, he had decided symptoms of congestion in the head, such as precede apoplexy. I questioned him as to his appetite. He informed me, that when he ate meat he had such an unconquerable desire for food about eleven o'clock, that he could not wait till noon. This he calls "meat hunger," for it disappeared soon after he came to the present style of living. He has no craving now; but when he begins to eat, the zest is exquisite.

Yours,

JOSHUA PORTER.

LETTER IV.—FROM DR. N. J. KNIGHT, OF TRURO.

Dated at TRURO, October, 1837.

DR. ALCOTT: SIR,—I hasten to comply so far with your request as to show my decided approbation of a fruit and farinaceous diet, both in health and sickness. The manner in which nutritious vegetables are presented to us for our consumption and support, evince to a demonstration the simplicity of our corporeal systems. Through every medium of correct information, we learn that the most distinguished men, both in ancient and modern times, were pre-eminently distinguished for their abstemiousness, and the simplicity of their diet.

It was not, however, a consideration of this kind that first induced me to relinquish flesh meat and fish. Some three years previous to my forming a determination to subsist upon farinacea, I had been laboring under an aggravated case of dyspepsia; and about six months previous, also, an attack of acute rheumatism.

I was harassed with constant constipation of the bowels, and ejection of food after eating, together with occasional pain in the head.

Under all these circumstances, I came to this determination, which I committed to paper: "November 9, 1831. This day ceased from strengthening this mortal body by any part of that which ever drew breath." To the above I rigidly adhered until last November, when my health had become so perfect that I thought myself invincible, so far as disease was concerned. All pains and aches had left me, and all the functions of the body seemed to be performed in a healthy manner.

My diet had consisted of rye and Indian bread, stale flour bread, sweet bread without shortening, milk, some ripe fruit, and occasionally a little butter.

During this time, while I devoted myself to considerable laborious practice and hard study, there was no deficiency of muscular strength or mental energy. I am fully satisfied my mind was never so active and strong.

Since last November I have, at times, taken animal food, in order that I might be absolutely satisfied that my mode of living acted decidedly in favor of my perfect health, and that a different course would produce organic derangement.

I had only taken animal food about two months after the usual custom, before I had a severe attack, and only escaped an inflammatory fever by the most rigid antiphlogistic treatment.

I again lived as I ought, and felt well; and having continued so some time, I resorted the second time to an animal diet.

In two months' time, I was taken with the urticaria febrilis, of Bateman, which lasted me more than two weeks, and my suffering was sufficient to forever exclude from my stomach every kind of animal food.

I am now satisfied, to all intents and purposes, that mankind would live longer, and enjoy more perfectly the "sane mind in a sound body," should they never taste flesh meat or fish.

A simple farinaceous diet I have ever found more efficient in the cure of chronic complaints, where there was not much organic lesion, than every other medical agent.

Mrs. A., infected with scrofula of the left breast, and in a state of ulceration, applied to me two years since. The ulcer was then the size of a half-dollar, and discharged a considerable quantity of imperfect pus. The axillary glands were much enlarged, and, doubting the practicability of operating with the knife in such cases, I told her the danger of her disease, and ordered her to subsist upon bread and milk and some fruit, drink water, and keep the body of as uniform temperature as possible. I ordered the sore to be kept clean by ablutions of tepid water. In less than three months, the ulcer was all healed, and her general health much improved. The axillary glands are still enlarged, though less so than formerly.

She still lives simply, and enjoys good health; but she tells me if she tastes flesh meat, it produces a twinging in the breast.

Many cases, like the above, have come under my observation and immediate attention, and suffice it to say, I have never failed to ameliorate the condition of every individual that has applied to me, who was suffering under chronic affections, if they would follow my prescriptions—unless the system was incapable of reaction.

Yours, truly,

N. J. KNIGHT.

LETTER V.—FROM DR. LESTER KEEP.

FAIR HAVEN, Jan. 22, 1838.

Dear Sir,—Agreeably to your request, I will inform you that from September, 1834, to June, 1836, I used no meat at all, except occasionally in my intercourse with society, I used a little to avoid attracting notice.

When I commenced my studies, life was burdensome. I knew not, for months, and I may say years, what enjoyment comfortable health affords. In a great many ways I can now see that I very greatly erred in my course of living. I am surprised that the system will hold out in its powers during so long a process in the use of what I should now consider the means best calculated to break it down.

I cannot now particularize. But in college, and during my professional studies, and since, during six or eight years of practice in an arduous profession, I have been greatly guilty, and neglected those means best calculated to promote and preserve health; and used those means best fitted to destroy it. The summers of 1832, 1833, and 1834, were pretty much lost, from wretched health. I was growing worse every year, and no medicines that I could prepare for myself, or that were prescribed by various brother physicians, had any thing more than a temporary effect to relieve me. All of the year 1834, until September, I used opium for relief; and I used three and four grains of sulphate of morphine per day, equal to about sixteen grains of opium. Spirit, wine, and ale I had tried, and journeys through many portions of the State of Maine, with the hope that a more northern climate would invigorate and restore a system that I feared was broken down forever, and that at the age of thirty-seven. But, without further preamble, I will say, I omitted at once and entirely the use of tea, coffee, meat, butter, grease of all sorts, cakes, pies, etc., wine, cider, spirits, opium (which I feared I must use as long as I lived), and tobacco, the use of which I learned in college. Of course, from so sudden and so great a change, a most horrid condition must ensue for many days, for the relief of which I used the warm bath at first several times a day. I had set no time to omit these articles, and made no resolutions, except to give this course a trial, to find out whether I had many native powers of system left, and what was their character and condition when unaffected by the list of agents mentioned.

I pursued this plan of living faithfully for one year and a half, and with unspeakable joy I found a gradual return of original vigor and health. Now, I cannot say that the omission of meat of all kinds, for a year and a half, caused this improvement in health; it is possible that it had but little to do with it. I know I was guilty of many bad habits; and probably all combined caused my bad condition.

At the close of the year and a half, I married my present second wife, and then commenced living as do others, in most respects, and continued this course most of the time until I received your letter. I then again omitted the use of all animal food, tea, coffee, and tobacco; and for the last month, it is a clear case, my health is better; that is, more vigorous to bear cold. I also bear labor and care better.

I have not investigated the subject of dietetics very much, but I have no doubt that the inhabitants of our whole land make too much use of animal food. No doubt it obstructs the vital powers, and tends to unbalance the healthful play and harmony of the various organs and their functions. There is too much nutriment in a small space. An unexpected quantity is taken; for with most people a sense of fullness is the test of a sufficient quantity.

I am satisfied that I am better without animal food than with the quantity I ordinarily use. If I should use but a small quantity once or twice a day, it is possible it would not be injurious. This I have not tried; for I am so excessively fond of meat, that I always eat *more* than a small quantity, when I eat it at all. Healthy, vigorous men, day laborers in the field, or forest, may perhaps require some meat to sustain the system, during hard and exhausting labor. Of this I cannot say.

I am now pretty well convinced, from two or three years' observation, that a large portion of my business, as a physician, arises from intemperance in the use of food. Too much and too rich nutriment is used, and my constant business is, to counteract its bad effects.

Two cases are now in mind of the great benefit of dieting for the recovery of health, the particulars of which I cannot now give you. One of them I think would be willing to speak for himself on the subject.

I am, sir, yours, etc.,

LESTER KEEP.

LETTER VI.—SECOND LETTER FROM DR. KEEP.

FAIR HAVEN, Ct., Jan. 26, 1838.

SIR,—Since I wrote you, a few days ago, I have learned of several individuals who have, for some length of time, used no flesh meat at all.

Amos Townsend, Cashier of the New Haven Bank, has, as I am told, lived almost entirely upon bread, crackers, or something of that kind, and but little of that. He can dictate a letter, count money, and hold conversation with an individual, all at the same time, with no embarrassment; and I know him to have firm health.

Our minister, Rev. B. L. Swan, during the whole of two years of his theological studies at Princeton, made crackers and water his only food, and was in good health.

Mr. Hanover Bradley, of this village, who has been several years a missionary among the Indians, has, for I think, eight or ten years, lived entirely on vegetable food. He had been long a dyspeptic.

There are some other cases of less importance, and probably very many in New Haven; but I am situated a mile from the city, and have never inquired for vegetable livers.

Yours, etc.,

LESTER KEEP.

LETTER VII.—FROM DR. HENRY H. BROWN

WEST RANDOLPH, Vt., Feb. 3, 1838.

DEAR SIR,—It has been about two years and a half since I adopted an exclusively vegetable diet, with no drink but water; and my food has been chiefly prepared by the most simple forms of cookery. Previously to this, I used a large proportion of flesh meat, and drank tea and coffee. I had much impaired my health by such indulgences. I hardly need to say that my health has greatly improved, and is now quite good and uniform.

I think that physicians, in prescribing for the removal of disease, should pay much more regard to the diet of their patients, and administer less of powerful medicine, than is customary with gentlemen of this profession at large.

Yours, etc.,

HENRY H. BROWN.

LETTER VIII.—FROM DR. FRANKLIN KNOX.

KINSTON,[5] N. C., June 23, 1837.

DEAR SIR,—Your letter of the 22d July has been hitherto unanswered, through press of business.

I consider an exclusive vegetable diet as of the utmost consequence in most diseases, especially in those chronic affections or morbid states of the system which are not commonly considered as diseases; and I think that, in these cases, such a diet is too often overlooked, even by physicians.

Yours, truly,

F. KNOX.

LETTER IX.—FROM A HIGHLY RESPECTABLE PHYSICIAN.

[The following letter, received last autumn, is from a medical gentleman, in a distant part of the country, whose name, for particular reasons, we stand pledged not to give to the world. The facts, however, may be relied on; and they are exceedingly important and interesting.]

DEAR SIR,—Your letter was duly received. I proceed to say that, since I settled in this town, my attacks of epilepsy[6] have occurred in the following order:

1833

Nov.	18.	One at	11 P. M.	Severe.
"	19.	"	"	"
"	24.	Nineteen, from	4 A. M. to 3 P. M.	Frightful.

1835.

Jan.	13.	One at	4 A. M.	}
"	15.	"	"	} Milder.
"	16.	Two at 2 and	4 A. M.	}

Thus it appears that I have enjoyed a longer immunity since the last, than for some years prior. I have maintained total abstinence from flesh, fish, or fowl, for two and a half years, namely, from March 1835 to the present time. That this happy immunity from a most obstinate disease is to be attributed solely to my abstinence from animal food, I do not feel prepared to assert; but that my general health has been better, my attacks of disease far milder, my vigor of mind and body greater, my mental perceptions clearer and more acute, and my enjoyment of life, on the whole, very essentially increased, I am fully prepared to prove.

I have, however, found it nearly as essential for me to abstain from many kinds of vegetable food as from animal, namely, from all kinds of flatulent vegetables; from all kinds of fruits and berries, except the very mildest—as, perfectly ripe and well baked sweet apples—and from all kinds of pies, sauces, and preserves. Of these, however, I am not able to say, as I do of the animal varieties, that I have practiced total abstinence; by no means. I have often ventured to indulge, and generally suffer more or less for my temerity. My severest sufferings for the last two years have been in the form of colic, of which I have had frequent slight attacks; but none to confine me over twenty-four hours.

ADDITIONAL STATEMENTS.—BY THE AUTHOR.[7]

From the age of five or six months to that of two years, I was literally crammed with flesh meat; usually of the most gross kind. Such a course was believed, by the fond parents and others, as likely to be productive of the most healthful and happy consequences. The result was an accumulation of adipose substance, that rendered me one of the most unsightly, not to say monstrous productions of nature. I ought not to say *nature*, perhaps; for, if not perverted, she produces no such monsters. At the age of six months, my weight was twenty-five pounds; and it rose soon after to thirty or more.

When I was about two years of age, I had the whooping-cough, and, having been brought up to the height, and more than the height of my condition, by over-feeding with fat meat, I suffered exceedingly. I? recovered, at length, but I had lost my relish, as I am informed, for flesh meat; and from this time till the age of fourteen, I seldom ate any but the leanest muscle. I was tolerably healthy, but, from the age of two years, was slender; so much so that, at five or six, I only weighed fifty pounds; and was constantly either found fault with, or pitied, because I did not eat meat in quality and quantity like other people. Nor was it without much effort, even at the age of fourteen, that I could bring myself to be reconciled to it. I was also trained to the early use of much cider, and to the moderate use of tea and spirits. I have spoken of my slender constitution;—I believe this was in part the result of excessive early labor, and that it was not wholly owing to a premature use of flesh meat.

I had suffered so much, however, from the belief that I was feeble from the latter cause, that I had no sooner become reconciled to the use of flesh and fish—which was at the age of fourteen—than I indulged in it quite freely. About this time I had a severe attack of measles, which came very near carrying me off. I was left with anasarca, or general dropsy, and with weak eyes. To cure the former the physicians plied me, for a long time, with blue pill, and with mercurial medicine in other forms, and also with digitalis; and finally filled my stomach to overflowing with diuretic drinks. However, in spite of them all, I recovered during the next year; except that a foundation was laid for premature decay of the teeth, and for a severe eruptive disease. This last, and the weakness of the eyes, were, for some time, very troublesome.

The eruptive complaint was soon discovered to be less severe, even in hot weather, and while I was using a great deal of exercise, in proportion as I abstained from all drinks but water, and ate none but mild food. Owing to the discovery of this fact and to other causes, I chiefly discontinued the use of stimulating food and drink, during the hottest part of the season; though I committed much error in regard to the quantity of my food, and drank quite too freely of cold water. Still I always found my health best, and my body and mind most vigorous at the end of summer, or the beginning of autumn, notwithstanding the very hard labor to which I was subjected on the farm. This increase of vigor was, at that time, attributed chiefly to a free use of summer fruits; for, so deeply had the belief been infixed by early education, that highly stimulating food and drink were indispensable to the full health and strength of mankind, and especially to people who were laboring hard, that, though I sometimes suspected they were not true friends to the human system, my conscience always condemned the suspicion, and pronounced me guilty of a species of high treason for harboring it.

This brings up my dietetic history, to the period at which it commences, in the letter to Dr. North. The study of medicine, however, from the age of twenty-four to twenty-seven, and the subsequent study and practice of it for a few years, joined to the changes I made at the same time in my physical habits, and my observations on their effects, led me to reject, one after another, and one group after another, the whole tribe of extra stimulants—solid and fluid.

The sequel of my story remains to be told. It is now nearly fifteen years since I wrote the letter, which is found at page 23d, to Dr. North. During this long period, and for several years before, amounting, in all, to about nineteen years, I have not only abstained entirely from flesh, fish, and fowl—not having eaten a pound of any one of these during the whole time, except the very few pounds I used in the time of the first visitation of our country with cholera, as before mentioned—but I have almost entirely abstained from butter, cheese, eggs, and milk. Butter, especially, I *never* taste at all. The occasional use of milk, in very small quantities, once a day, has, however, been resorted to; not from necessity, indeed, or to gratify any strong desire or inclination for it, but from a conviction of its happy medicinal effects on my much-injured frame. Hot food of every kind, and liquids, with the exception just made, I rarely touch. Nearly every thing is taken in as solid a form and in as simple a state as possible; with no condiments, except a very little salt, and with no sweets, sauces, gravies, jellies, preserves, etc. I seldom use more than one sort of food at a time, unless it be to add fruit as a second article; and this is rarely done, except in the morning. I have for ten or twelve years used no drinks with my meals; and sometimes for months together have had very little thirst at all.[8]

And as to the effects, they are such, and have all along been such, as to make me wonder at myself, whenever I think of it. Instead of being constantly subject to cold, and nearly dying with consumption in the spring, I am almost free from any tendency to take cold at all. During the winter of 1837-8, by neglecting to keep the temperature of my room low enough, and by neglecting also to take sufficient exercise in the open air, I became unusually tender, and suffered to some extent from colds. But I was well again during the spring, and felt as if I had recovered or nearly recovered my former hardihood.

In regard to other complaints, I may say still more. Of rheumatism, I have scarcely had a twinge in twelve or fourteen years. My eruptive complaint is, I believe, *entirely* gone. The weakness of my eyes has been wholly gone for many years. Indeed, the strength and perfection of my sight and of all my senses, till nearly fifty years of age—hearing perhaps excepted, in which I perceive no alteration—appeared to be constantly improving. My stomach and intestines perform their respective duties in the most appropriate, correct, and healthful manner. My appetite is constantly good, and as constantly improving;—that is, going on toward perfection. I can detect, especially by taste, almost any thing which is in the least offensive or deleterious in food or drink; and yet I can receive, without immediate apparent disturbance, and readily digest, almost any thing which ever entered a human stomach—knives, pencils, clay, chalk, etc., perhaps excepted. I can eat a full meal of cabbage, or any other very objectionable crude aliment, or even cheese or pastry—a single meal, I mean—with apparent impunity; not when fatigued, of course, or in any way debilitated, but in the morning and when in full strength. It is true, I make no experiments of this sort, except occasionally *as* experiments.

In my former statements I gave it as my opinion that vegetable food was less aperient than animal. My opinion now is, that if we were trained on vegetable food, and had never received substances into the stomach which were unduly stimulating, we should find the intestinal or peristaltic action quite sufficient. The apparent sluggishness of the bowels, when we first exchange an animal diet for a vegetable one, is probably owing to our former abuses. At present, I find my plain vegetable food, in moderate and reasonable quantity, quite as aperient as it ought to be, and, if I exceed a proper quantity, too much so.

I have now no remaining doubts of the vast importance that would result to mankind, from an universal training from childhood, to the exclusive use of vegetable food. I believe such a course of training, along with a due attention to air, exercise, cleanliness, etc., would be the means of improving our race, physically, intellectually, and morally, beyond any thing of which the world has yet conceived. But my reasons for this belief will be seen more fully in another place. They are founded in science and the observation of facts around me, much more than on a narrow individual experience.

There is one circumstance which I must not omit, because it is full of admonition and instruction. I have elsewhere stated that, twenty-three years ago, I had incipient phthisis. Of this fact, and of the fact that there were considerable inroads made by disease on the upper lobe of the right lung, I have not the slightest doubt. The symptoms were such at the time, and subsequently, as could not have been mistaken. Besides, what was, as I conceive, pretty fully established by the symptoms which existed, is rendered still more certain by auscultation. The sounds which are heard during respiration, in the region to which I have alluded, leave no doubt on the minds of skillful medical men, of their origin. Still I doubt whether the disease has made any considerable progress for many years.

But, during the winter of 1837-8, my employments became excessively laborious; and, for the whole winter and spring, were sufficient for at least two healthy and strong men. They were also almost wholly sedentary. At the end of May, I took a long and rather fatiguing journey through a country by no means the most healthy, and came home somewhat depressed in mind and body, especially the former. I was also unusually emaciated, and I began to have fears of a decline. Still, however, my appetite was good, and I had a good share of bodily strength. The more I directed my attention to myself, the worse I became; and I actually soon began to experience darting pains in the chest, together with other symptoms of a renewal of pulmonary disease. Perceiving my danger, however, from the state of my mind, I at length made a powerful effort to shake off the mental disturbance—which succeeded. This, together with moderate labor and rather more exercise than before, seemed gradually to set me right.

Again, in the spring of 1848, after lecturing for weeks and months—often in bad and unventilated rooms and subjecting myself, unavoidably, to many of those abuses which exist every where in society, I was attacked with a cough, followed by great debility, from which it cost me some three months or more of labor with the spade and hoe, to recover. With this and the exceptions before named, I have now, for about twenty years, been as healthy as ever I was in my life, except the slight tendency to cold during the winter of which I have already taken notice. I never was more cheerful or more happy; never saw the world in a brighter aspect; never before was it more truly "morning all day" with me. I have paid, in part, the penalty of my transgressions; and may, perhaps, go on, in life, many years longer.

I now fear nothing in the future, so far as health and disease are concerned, so much as excessive alimentation. To this evil—and it is a most serious and common one in this land of abundance and busy activity—I am much exposed, both from the keenness of my appetite, and the exceeding richness of the simple vegetables and fruits of which I partake. But, within a few years past, I seem to have gotten the victory, in a good measure, even in this respect. By eating only a few simple dishes at a time, and by measuring or weighing them with the eye—for I weigh them in no other way—I am usually able to confine myself to nearly the proper limits.

This caution, and these efforts at self-government, are not needed because their neglect involves any immediate suffering; for, as I have already stated, there was never a period in my life before, when I was so completely independent—apparently so, I mean—of external circumstances. I can eat what I please, and as much or as little as I please. I can observe set hours, or be very irregular. I can use a pretty extensive variety at the same meal, and a still greater variety at different meals, or I can live perpetually on a single article—nay, on almost any thing which could be named in the animal or vegetable kingdom—and be perfectly contented and happy in the use of it. I could in short, eat, work, think, sleep, converse, or play almost all the while; or I could abstain from any or all of these, almost all the while. Let me be understood, however. I do not mean to say that either of these courses would be best for me, in the end; but only that I have so far attained to independence of external circumstances that, for a time, I believe I should be able to do or bear all I have mentioned.

One thing more, in this connection, and I shall have finished my remarks. I sleep too little; but it is because I allow my mind to run over the world so much, and lay so many schemes for human improvement or for human happiness; and because I allow my sympathies to become so deeply enlisted in human suffering and human woe. I should be most healthy, in the end, by spending six hours or more in sleep; whereas I do not probably exceed four or five. I have indeed obtained a respite from the grave of twenty-three years, through a partial repentance and amendment of life, and the mercy of God; but did I obey all his laws as well as I do a part of them, I know of no reason why my life might not be lengthened, not merely fifteen years, as was Hezekiah's, or twenty-three merely, but forty or fifty.

FOOTNOTES:

[5] Dr. Knox has since removed to St. Louis, Missouri.

[6] The reader will find another remarkable cure of epilepsy in a subsequent chapter of this volume. The case was that of Dr. Taylor, of England.

[7] See pages 13 and 23.

[8] This fact, and certain discussions on the subject of temperance, led me to abstain, about the years 1841 and 1842, entirely from all drink for a long

time. Indeed, I made two of these experiments; in one of which I abstained nine months and nineteen days, and in the other fourteen months and one or two days; except that in the latter case I ate, literally, for one or two successive days, while working hard at haying, one or two bowls a day of bread and water. But these were experiments *merely*—the experiments made by a medical man who preferred making experiments on himself to making them on others; and they never deserved the misconstruction which was put upon them by several persons, who, in other respects, were very sensible men. "The author" never believed with Dr. Lambe, of London, that man is not a drinking animal.

CHAPTER V.

TESTIMONY OF OTHER MEDICAL MEN, BOTH OF ANCIENT AND MODERN TIMES.

General Remarks.—Testimony of Dr. Cheyne.—Dr. Geoffroy.—Vanquelin and Percy.—Dr. Pemberton.—Sir John Sinclair.—Dr. James.—Dr. Cranstoun.—Dr. Taylor.—Drs. Hufeland and Abernethy.—Sir Gilbert Blane.—Dr. Gregory.—Dr. Cullen.—Dr. Rush.—Dr. Lambe.—Prof. Lawrence.—Dr. Salgues.—Author of "Sure Methods."—Baron Cuvier.—Dr. Luther V. Bell.—Dr. Buchan.—Dr. Whitlaw.—Dr. Clark.—Prof. Mussey.—Drs. Bell and Condie.—Dr. J. V. C. Smith.—Mr. Graham.—Dr. J. M. Andrews, Jr.—Dr. Sweetser.—Dr. Pierson.—Physician in New York.—Females' Encyclopedia.—Dr. Van Cooth.—Dr. Beaumont.—Sir Everard Home.—Dr. Jennings.—Dr. Jarvis.—Dr. Ticknor.—Dr. Coles.—Dr. Shew.—Dr. Morrill.—Dr. Bell.—Dr. Jackson.—Dr. Stephenson.—Dr. J. Burdell.—Dr. Smethurst.—Dr. Schlemmer.—Dr. Curtis.—Dr. Porter.

GENERAL REMARKS.

The number of physicians, and surgeons, and medical men, whose testimony is brought to bear on the subject of diet, in the chapter which follows, is by no means as great as it might have been. There are few writers on anatomy, physiology, materia medica, or disease, who have not, either directly or indirectly, given their testimony in favor of a mild and vegetable diet for persons affected with certain chronic diseases. And there is scarcely a writer on hygiene, or even on diet, who has not done much more than this, and at times hinted at the safety of such a diet for those who are in health; particularly the studious and sedentary. But my object has been, not so much to collect all the evidence I could, as to make a judicious selection—a selection which should present the subject upon which it bears, in as many aspects as possible. I have aimed in general, also, to procure the testimony of intelligent and philanthropic men; or, at least of men whose names have by some means or other been already brought before the public. If there are a few exceptions to this rule, if a few are men whose names have been hitherto unknown, it is on account of the *aspect*, as I have already said, of their testimony, or on account of their peculiar position, as regards country, age of the world, etc., or to secure their authority for certain anecdotes or facts.

In the arrangement of the testimony, I have been guided by no particular rule, unless it has been to present first that of some of the older and most accredited writers, such as Cheyne, Cullen, and Rush. The testimony of certain living men and authors, particularly of our own country, has been presented toward the close of the chapter, and in a very brief and condensed form, from design. The propriety of inserting their names at all was for a time considered doubtful. It is believed, however, that they could not, in strict justice, have been entirely omitted. But let not the meagre sketch of their views I have given, satisfy us. We want a full development of their principles from their own pens—such a development as, I hope, will not long be withheld from a world which is famishing for the want of it. But now to the testimony.

DR. GEORGE CHEYNE.

This distinguished physician, and somewhat voluminous writer, flourished more than a hundred years ago. He may justly be esteemed the father of what is now called the "vegetable system" of living; although it is evident he did not see every thing clearly. "In the early part of his life," says Prof. Hitchcock, in his work on Dyspepsia, "he was a voluptuary; and before he attained to middle age, was so corpulent that it was necessary to open the whole side of his carriage that he might enter; and he saw death inevitable, without a change of his course. He immediately abandoned all ardent spirits, wine, and fermented liquors, and confined himself wholly to milk, vegetables, and water. This course, with active exercise, reduced him from the enormous weight of four hundred and forty-eight pounds, to one hundred and forty; and restored his health and the vigor of his mind. After a few years, he ventured to change his abstemious diet for one more rich and stimulating. But the effect was a recurrence of his former corpulence and ill health. A return to milk, water, and vegetables restored him again; and he continued in uninterrupted health to the age of seventy-two."

The following is his account of himself, at the age of about seventy:

"It is now about sixteen years since, for the last time, I entered upon a milk and vegetable diet. At the beginning of this period, I took this light food as my appetite directed, without any measure, and found myself easy under it. After some time, I found it became necessary to lessen the quantity; and I have latterly reduced it to one half, at most, of what I at first seemed to bear. And if it shall please God to spare me a few years longer, in order, in that case, to preserve that freedom and clearness which, by his, blessing, I now enjoy, I shall probably find myself obliged to deny myself one half of my present daily substance—which is precisely three Winchester pints of new cows' milk, and six ounces of biscuit made of fine flour, without salt or yeast, and baked in a quick oven."

It is exceedingly interesting to find an aged physician, especially one who had formerly been in the habit of using six pints of milk, and twelve ounces of unfermented biscuit, and of regarding that as a low diet, reducing himself to one half this quantity in his old age, with evident advantages; and cheerfully looking forward to a period, as not many years distant, when he should be obliged to restrict himself to half even of that quantity. How far he finally carried his temperance, we do not exactly know. We only know that, after thirty years of health and successful medical practice, he strenuously contended for the superiority of a vegetable and milk diet over any other, whether for the feeble or the healthy. But his numerous works abound with the most earnest exhortations to temperance in all things, and with the most interesting facts and cogent reasonings; and—I repeat it—if there be any individual, since the days of Pythagoras, whose name ought to be handed down to posterity as the father of the vegetable system of living, it is that of Dr. Cheyne.

Among his works are, a work on Fevers; an Essay on the true Nature and proper Method of treating the Gout; a work on the Philosophical Principles of Religion; an Essay of Health and Long Life; a work called the English Malady; and another entitled the Natural Method of Cure in the Diseases of the Body, and the Distempers of the Mind depending thereon. The latter, and his Essay of Long Life are, in my view, his greatest works; though the history of his own experience is chiefly contained in his English Malady.

I shall now proceed to make such extracts from his works, as seem to me most striking and important to the general reader. They are somewhat numerous, and there may be a few repetitions; but I was more anxious to preserve his exact language—which is rather prolix—than to abridge too much, at the risk of misrepresenting his sentiments.

"When I see milk, oil, emulsion, mild watery fluids, and such like soft liquors run through leathern tubes or pipes (for such animal veins and arteries indeed are) for years, without destroying them, and observe on the other hand that brine, inflammable or urinous spirits, and the like acrimonious and burning fluids corrode, destroy, and consume them in a very short time; when I consider the rending, burning, and tearing pains and tortures of the gout, stone, colic, cancer, rheumatism, convulsions, and such like insufferably painful distempers; when I see the crises of almost all acute distempers happen either by rank and fetid sweats, thick lateritious and lixivious sediments in the urine, black, putrid, and fetid dejections, attended with livid and purple spots, corrosive ulcers, impostumes in the joints or muscles, or a gangrene and mortification in this or that part of the body; when I see the sharp, the corroding and burning ichor of scorbutic and scrofulous sores, fretting, galling, and blistering the adjacent parts, with the inflammation, swelling, hardness, scabs, scurf, scales, and other loathsome cutaneous foulnesses that attend, the white gritty and chalky matter, and hard stony or flinty concretions which happen to all those long troubled with severe gouts, gravel, jaundice, or colic—the obstructions and hardnesses, the putrefaction and mortification that happen in the bowels, joints, and members in some of these diseases, and the rottenness in the bones, ligaments, and membranes that happen in others; all the various train of pains, miseries, and torments that can afflict any part of the compound, and for which there is scarce any reprieve to be obtained, but by swallowing a kind of poison (opiates, etc.); when I behold with compassion and sorrow, such scenes of misery and woe, and see them happen only to the rich, the lazy, the luxurious, and the inactive, those who fare daintily and live voluptuously, those who are furnished with the rarest delicacies, the richest foods, and the most generous wines, such as can provoke the appetites, senses, and passions, in the most exquisite and voluptuous manner; to those who leave no desire or degree of appetite unsatisfied, and not to the poor, the low, the meaner sort, those destitute of the necessaries, conveniences, and pleasures of life; to the frugal, industrious, temperate, laborious, and active, inhabiting barren and uncultivated countries, deserts, and forests under the poles or under the line;—I must, if I am not resolved to resist the strongest conviction, conclude that it must be something received into the body that can produce such terrible appearances in it—some flagrant and notable difference in the food that so sensibly distinguishes them from the latter; and that it is the miserable man himself that creates his miseries and begets his torture, or at least those from whom he has derived his bodily organs.

"Nothing is so light and easy to the stomach, most certainly, as the farinaceous or mealy vegetables; such as peas, beans, millet, oats, barley, rye, wheat, sago, rice, potatoes, and the like."

Milk is not included in the foregoing list of light articles; although Dr. C. was evidently extremely fond of prescribing it in chronic diseases. It does not fully appear, so far as I can learn from his writings, that he regarded it as by any means indispensable to those who were perfectly healthy, except during infancy and childhood. The following extract will give us—more than any other, perhaps—his real sentiments, though modestly expressed in the form of a conjecture, rather than a settled belief.

"I have sometimes indulged the conjecture that animal food, and *made* or artificial liquors, in the original frame of our nature and design of our creation, were not intended for human creatures. They seem to me neither to have those strong and fit organs for digesting them (at least, such as birds and beasts of prey have that live on flesh); nor, naturally, to have those voracious and brutish appetites, that require animal food and strong liquors to satisfy them; nor those cruel and hard hearts, or those diabolical passions, which could easily suffer them to tear and destroy their fellow-creatures; at least, not in the first and early ages, before every man had corrupted his way, and God was forced to exterminate the whole race by an universal deluge, and was also obliged to shorten their lives from nine hundred or one thousand years to seventy. He wisely foresaw that animal food and artificial liquors would naturally contribute toward this end, and indulged or permitted the generation that was to plant the earth again after the flood the use of them for food; knowing that, though it would shorten their lives and plait a scourge of thorns for the backs of the lazy and voluptuous, it would be cautiously avoided by those who knew it was their duty and happiness to keep their passions low, and their appetites in subjection. And this very era of the flood is that mentioned in holy writ for the indulgence of animal food and artificial liquors, after the trial had been made how insufficient alone a vegetable diet—which was the first food appointed for human kind after their creation—was, in the long lives of men, to restrain their wickedness and malice, and after finding that nothing but shortening their duration could possibly prevent the evil.

"It is true, there is scarce a possibility of preventing the destroying of animal life, as things are now constituted, since insects breed and nestle in the very vegetables themselves; and we scarcely ever devour a plant or root, wherein we do not destroy innumerable animalculæ. But, besides what I have said of nature's being quite altered and changed from what was originally intended, there is a great difference between destroying and extinguishing animal life by choice and election, to gratify our appetites, and indulge concupiscence, and the casual and unavoidable crushing of those who, perhaps, otherwise would die within the day, or at most the year, and who obtain but an inferior kind of existence and life, at the best.

"Whatever there may be, in this conjecture, it is evident to those who understand the animal economy of the frame of human bodies, together with the history, both of those who have lived abstemiously, and of those who have lived freely, that indulging in flesh meat and strong liquors, inflames the passions and shortens life, begets chronic distempers and a decrepit age.

"For remedying the distempers of the body, to make a man live as long as his original frame was designed to last, with the least pain and fewest diseases, and without the loss of his senses, I think Pythagoras and Cornaro by far the two greatest men that ever were:—the first, by vegetable food and unfermented liquors; the latter, by the lightest and least of animal food, and naturally fermented liquors. Both lived to a great age. But, what is chiefly to be regarded in their conduct and example, both preserved their senses, cheerfulness, and serenity to the last; and, which is still more to be regarded, both, at least the last, dissolved without pain or struggle; the first having lost his life in a tumult, as it is said by some, after a great age of perfect health.

"A plain, natural, and philosophical reason why vegetable food is preferable to all other food is, that abounding with few or no salts, being soft and cool, and consisting of parts that are easily divided and formed into chyle without giving any labor to the digestive powers, it has not that force to open the lacteals, to distend their orifices and excite them to an unnatural activity, to let them pass too great a quantity of hot and rank chyle into the blood, and so overcharge and inflame the lymphatics and capillaries, which is the natural and ordinary effect of animal food; and therefore cannot so readily produce diseases. There is not a sufficient stimulus in the salts and spirits of vegetable food to create an unnatural appetite, or violent cramming; at least, not sufficient to force open and extend the mouths of the lacteals, more than naturally they are or ought to be. Such food requires little or no force of digestion, a little gentle heat and motion being sufficient to dissolve it into its integral particles: so that, in a vegetable diet, though the sharp humors, in the first passages, are extended, relaxed stomach, and sometimes a delightful piquancy in the food, may tempt one to exceed in quantity; yet rarely, if spices and sauces—as too much butter, oil, and sugar—are not joined to seeds[9] and vegetables, can the mischief go farther than the stomach and bowels, to create a pressed load, sickness, vomiting, or purging, by its acquiring an acrimony from its not being received into the lacteals;—so that on more being admitted into the blood than the expenses of living require, life and health can never be endangered by a vegetable diet. But all the contrary happens under a high animal diet."

Now I will not undertake to vouch—as indeed I cannot, conscientiously, do it—for the correctness of all Dr. C.'s notions in physiology or pathology. The great object I have in view, by the introduction of these quotations, may be accomplished without it. His preference for vegetable food, or for what he calls a milk and seed diet, is the point which I wish to make most prominent.

In the following paragraphs, he takes up and considers some of the popular objections of the day, to his doctrines and practice.

"One of the most terrible objections some weak persons make against this regimen and method, is, that upon accidental trials, they have always found milk, fruit, and vegetables so inflate, blow them up, and raise such tumults and tempests in their stomach and bowels, that they have been terrified and affrighted from going on. I own the truth and fact to be such, in some as is represented; and that in stomachs and entrails inured only to hot and high meats and drinks, and consequently in an inflammatory state and full of choler and phlegm, this sensation will sometimes happen—just as a bottle of cider or fretting wine, when the cork is pulled out, will fly up, and fume, and rage; and if you throw in a little ferment or acid (such as milk, seeds, fruit, and vegetables *to them*), the effervescence and tempest will exasperate to a hurricane.

"But what are wind, flatulence, phlegm, and choler? What, indeed, but stopped perspiration, superfluous nourishment, inconcocted chyle, of high food and strong liquors, fermented and putrifying? And when these are shut up and corked, with still more and more solid, strong, hot, and styptic meats and drinks, is the corruption and putrefaction thereby lessened? Will it not then, at last, either burst the vessel, or throw out the cork or stopples, and raise still more lasting and cruel tempests and tumults? Are milk and vegetables, seeds and fruits, harder of digestion, more corrosive, or more capable of producing chyle, blood, and juices, less fit to circulate, to perspire, and be secreted?

"But what is to be done? The cure is obvious. Begin by degrees; eat less animal food—the most tender and young—and drink less strong fermented liquors, for a month or two. Then proceed to a *trimming* diet, of one day, seed and vegetables, and another day, tender, young animal food;—and, by degrees, slide into a total milk, seed, and vegetable diet; cooling the stomach and entrails gradually, to fit them for this soft, mild, sweetening regimen; and in time your diet will give you all the gratification you ever had from strong, high, and rank food, and spirituous liquors. And you will, at last, enjoy ease, free spirits, perfect health, and long life into the bargain.

"Seeds of all kinds are fittest to begin with, in these cases, when dried, finely ground, and dressed; and, consequently, the least flatulent. Lessen the quantity, even of these, below what your appetite would require, at least for a time. Bear a little, and forbear.

"Virtue and good health are not to be obtained, without some labor and pains, against contrary habits. It was a wild bounce of a Pythagorean, who defied any one to produce an instance of a person, who had long lived on milk and vegetables, who ever cut his own throat, hanged, or made way with himself; who had ever suffered at Tyburn, gone to Newgate, or to Moorfields; (and, he added rather profanely,) or, would go to eternal misery hereafter.

"Another weighty objection against a vegetable diet, I have heard, has been made by learned men; and is, that vegetables require great labor, strong exercise, and much action, to digest and turn them into proper nutriment; as (say they) is evident from their being the common diet of day-laborers, handicraftsmen, and farmers. This objection I should have been ashamed to mention, but that I have heard it come from men of learning; and they might have as justly said, that freestone is harder than marble, and that the juice of vegetables makes stronger glue than that of fish and beef!

"Do not children and young persons, that is, tender persons, live on milk and seeds, even before they are capable of much labor and exercise? Do not all the eastern and southern people live almost entirely on them? The Asiatics, Moors, and Indians, whose climates incapacitate them for much labor, and whose indolence is so justly a reproach to them,—are these lazier and less laborious men than the Highlanders and native Irish?

"The truth is, hardness of digestion principally depends on the minuteness of the component particles, as is evident in marble and precious stones. And animal substances being made of particles that pass through innumerable very little, or infinitely small excretory ducts, must be of a much finer texture, and consequently harder, or tougher, in their composition, than any vegetable substance can be. And the flesh of animals that live on animals, is like double distilled spirits, and so requires much labor to break, grind, and digest it. And, indeed, if day-laborers, and handicraftsmen were allowed the high, strong food of men of condition, and the quiet and much-thinking persons were confined to the farmer and ploughman's food, it would be much happier for both.

"Another objection, still, against a milk and vegetable diet is, that it breeds phlegm, and so is unfit for tender persons, of cold constitutions; especially those whose predominant failing is too much phlegm. But this objection has as little foundation as either of the preceding. Phlegm is nothing but superfluous chyle and nourishment, as the taking down more food than the expenses of living and the waste of the solids and fluids require. The people that live most on such foods—the eastern and southern people and those of the northern I have mentioned—are less troubled with phlegm than any others. Superfluity will always produce redundancy, whether it be of phlegm or choler; and that which will digest the most readily, will produce the least phlegm—such as milk, seeds, and vegetables. By cooling and relaxing the solids, the phlegm will be more readily thrown up and discharged—more, I say, by such a diet than by a hot, high, caustic, and restringent one; but that discharge is a benefit to the constitution, and will help it the sooner and faster to become purified, and so to get into perfect good health. Whereas, by shutting them up, the can or cask must fly and burst so much the sooner.

"The only material and solid objections against a milk, seed, and vegetable diet, are the following:

"*First*, That it is particular and unsocial, in a country where the common diet is of another nature. But I am sure sickness, lowness, and oppression, are much more so. These difficulties, after all, happen only at first, while the cure is about; for, when good health comes, all these oddnesses and specialities will vanish, and then all the contrary to these will be the case.

"*Secondly*, That it is weakening, and gives a man less strength and force, than common diet. It is true that this may be the result, at first, while the cure is imperfect. But then the greater activity and gayety which will ensue on the return of health, under a milk and vegetable diet, will liberally supply that defect.

"*Thirdly*, The most material objection against such a diet is, that it cools, relaxes, softens, and unbends the solids, at first, faster than it corrects and sweetens the juices, and brings on greater degrees of lowness than it is designed to cure; and so sinks, instead of raising. But this objection is not universally true; for there are many I have treated, who, without any such inconvenience, or consequent lowness, have gone into this regimen, and have been free from any oppression, sinking, or any degree of weakness, ever after; and they were not only those who have been generally temperate and clean, free from humors and sharpnesses, but who, on the decline of life, or from a naturally weak constitution or frame, have been oppressed and sunk from their weakness and their incapacity to digest common animal food and fermented liquors.

"I very much question if any diet, either hot or cool, has any great influence on the solids, after the fluids have been entirely sweetened and balmified. Sweeten and thin the juices, and the rest will follow, as a matter of course."

At page 90 of Dr. Cheyne's Natural Method of Curing Diseases, he thus says:

"People think they cannot possibly subsist on a little meat, milk, and vegetables, or on any low diet, and that they must infallibly perish if they should be confined to water only; not considering that nine tenths of the whole mass of mankind are necessarily confined to this diet, or pretty nearly to it, and yet live with the use of their senses, limbs, and faculties, without diseases, or but few, and those from accidents or epidemical causes; and that there have been nations, and now are numbers of tribes, who voluntarily confine themselves to vegetables only; as the Essenes among the Jews, some Hermits and Solitaries among the Christians of the first ages, a great number of monks in the Chartreux now in Europe, Banians among the Indians and Chinese, the Guebres among the Persians, and of old, the Druids among ourselves."

To illustrate the foregoing, I may here introduce the following extracts from the sixth London edition of Dr. Cheyne's Essay on Health and Long Life.

"It is surprising to what a great age the Eastern Christians, who retired from the persecutions into the deserts of Egypt and Arabia, lived healthful on a very little food. We are informed, by Cassian, that the common measure for twenty-four hours was about twelve ounces, with only pure water for drink. St. Anthony lived to one hundred and five years on mere bread and water, adding only a few herbs at last. On a similar diet, James the Hermit lived to one hundred and four years. Arsenius, the tutor of the emperor Arcadius, to one hundred and twenty—sixty-five years in society, and fifty-five in the desert. St. Epiphanius, to one hundred and fifteen; St. Jerome, about one hundred; Simon Stylites, to one hundred and nine; and Romualdus, to one hundred and twenty.

"It is wonderful in what sprightliness, strength, activity, and freedom of spirits, a low diet, even here in England, will preserve those who have habituated themselves to it. Buchanan informs us of one Laurence, who preserved himself to one hundred and forty, by the mere force of temperance and labor. Spotswood mentions one Kentigern (afterward called St. Mongah, or Mungo, from whom the famous well in Wales is named), who lived to one hundred and eighty-five years; and who, after he came to years of understanding, never tasted wine or strong drink, and slept on the cold ground.

"My worthy friend, Mr. Webb, is still alive. He, by the quickness of the faculties of the mind, and the activity of the organs of his body, shows the great benefit of a low diet—living altogether on vegetable food and pure water. Henry Jenkins lived to one hundred and sixty-nine years on a low, coarse, and simple diet. Thomas Parr died at the age of one hundred and fifty-two years and nine months. His diet was coarse bread, milk, cheese, whey, and small beer; and his historian tells us, that he might have lived a good while longer if he had not changed his diet and air; coming out of a clear, thin air, into the thick air of London, and being taken into a splendid family, where he fed high, and drank plentifully of the best wines, and, as a necessary consequence, died in a short time. Dr. Lister mentions eight persons in the north of England, the youngest of whom was above one hundred years old, and the oldest was one hundred and forty. He says, it is to be observed that the food of all this mountainous country is exceeding coarse."

Dr. C., in his Natural Method, at page 91, thus continues his remarks:

"And there are whole villages in this kingdom, even of those who live on the plains, who scarce eat animal food, or drink fermented liquors a dozen times a year. It is true, most of these cannot be said to live at ease and commodiously, and many may be said to live in barbarity and ignorance. All I would infer from this is, that they do live, and enjoy life, health, and outward serenity, with few or no bodily diseases but from accidents and epidemical causes; and that, being reduced by voluntary and necessary poverty, they are not able to manage with care and caution the rest of the non-naturals, which, for perfect health and cheerfulness, must all be equally attended to, and prudently conducted; and their ignorance and brutality is owing to the want of the convenience of due and sufficient culture and education in their youth.

"But the only conclusion I would draw from these historical facts is, that a low diet, or living on vegetables, will not destroy life or health, or cause nervous and cephalic distempers; but, on the contrary, cure them, as far as they are curable. I pretend to demonstrate from these facts, that abstinence and a low diet is the great antidote and universal remedy of distempers acquired by excess, intemperance, and a mistaken regimen of high meats and drinks; and that it will greatly alleviate and render tolerable the original distempers derived from diseased parents; and that it is absolutely necessary for the deep thinking part of mankind, who would preserve their faculties sound and entire, ripe and pregnant to a green old age and to the last dregs of life; and that it is, lastly, the true and real antidote and preservative from heavy-headedness, irregular and disorderly intellectual functions, from loss of the rational faculties, memory, and senses, and from all nervous distempers, as far as the ends of Providence and the condition of mortality will allow.

"Let two people be taken as nearly alike as the diversity and the individuality of nature will admit, of the same age, stature, complexion, and strength of body, and under the same chronical distemper, and I am willing to take the seeming worse of the two; let all the most promising nostrums, drops, drugs, and medicines known among the learned and experienced physicians, ancient or modern, regular physicians or quacks, be administered to the best of the two, by any professor at home or abroad; I will manage my patient with only a few naturally indicated and proper evacuations and sweetening innocent alternatives, which shall neither be loathsome, various, nor complicated, require no confinement, under an appropriate diet, or, in a word, under the 'lightest and the least,' or at worst under a milk and seed diet; and I will venture reputation and life, that my method cures sooner, more perfectly and durably, is much more easily and pleasantly passed through, in a shorter time, and with less danger of a relapse than the other, with all the assistance of the best skill and experience, under a full and free, though even a commonly reputed moderate diet, but of rich foods and generous liquors; much more, under a voluptuous diet."

But I am unwilling to dismiss this subject without inserting a few more extracts from Dr. Cheyne, to show his views of the treatment of diseases. And first, of the scurvy, and other diseases which he supposes to arise from it.

"There is no chronical distemper, whatsoever, more universal, more obstinate, and more fatal in Britain than the scurvy, taken in its general extent. Scarce any one chronical distemper but owes its origin to a scorbutic tendency, or is so complicated with it, that it furnishes the most cruel and most obstinate symptoms. To it we owe all the dropsies that happen after the meridian of life; all diabetes, asthmas, consumptions of several kinds; many sorts of colics and diarrhœas; some kinds of gouts and rheumatisms, all palsies, various kinds of ulcers, and possibly the cancer itself; and most cutaneous foulnesses, weakly constitutions, and bad digestions; vapors, melancholy, and almost all nervous distempers whatsoever. And what a plentiful source of miseries the last are, the afflicted best can tell. And scarce any one chronical distemper whatever, but has some degree of this evil faithfully attending it. The reason why the scurvy is peculiar to this country and so fruitful of miseries, is, that it is produced by causes mostly special and particular to this island, to wit: the indulging so much in animal food and strong fermented liquors, sedentary and confined employments, etc.

"Though the inhabitants of Britain live, for the most part, as long as those of a warmer climate, and probably rather longer, yet scarce any one, especially those of the better sort, but becomes crazy and suffers under some chronical distemper or other, before he arrives at old age.

"Nothing less than a very moderate use of animal food, and that of the least exciting kind, and a more moderate use of spirituous liquors, due exercise, etc., can keep this hydra under. And nothing else than a total abstinence from animal food and alcoholic liquors can totally extirpate it."

The following are extracted from his "Natural Methods." I do not lay them down as recipes, to be followed in the treatment of diseases; but to show the views of Dr. Cheyne in regard to vegetable regimen.

"1. *Cancer.*—Any cancer that can be cut out, contracted, and healed up with common, that is, soft, cool, and gently astringent dressings, and at last left as an issue on the part, may, by a cow's milk and seed diet continued ever afterward, be made as easy to the patient, and his life and health as long preserved, almost, as if he had never been afflicted with it; especially if under fifty years of age.

"2. *Cancer.*—A total ass's milk diet—about two quarts a day, without any other meat or drink—will in time cure a cancer in any part of the body, with mere common dressings, provided the patient is not quite worn out with it before it is begun, or too far gone in the common duration of life and even in that case, it will lessen the pain, lengthen life, and make death easier, especially if joined with small interspersed bleedings, millepedes, crabs' eyes prepared, nitre and rhubarb, properly managed. But the diet, even after the cure, must be continued, and never after greatly altered, unless it be into cow's milk with seeds.

"3. *Consumption.*—A total milk and seed diet, gentle and frequent bleedings, as symptoms exasperate, a little ipecacuanha or thumb vomit repeated once or twice a week, chewing quill bark in the morning, and a few grains of rhubarb at night, will totally cure consumptions, even when attended with tubercles, and hemoptoe, and hectic, in the first stage; will greatly relieve, if not cure, in the second stage, especially if riding and a warm clear air be joined; and make death easier in the third and last stage.

"4. *Fits.*—A total cow's milk diet—about two quarts a day—without any other food, will at last totally cure all kinds of fits, epileptical, hysterical, or apoplectic, if entered upon before fifty. But the patient, if near fifty, must ever after continue in the same diet, with the addition only of seeds; otherwise his fits will return oftener and more severely, and at last cut him off.

"5. *Palsy.*—A total cow's milk diet, without any other food, will bid fairest to cure a hemiplegia or even a dead palsy, and consequently all the lesser degrees of a partial one, if entered upon before fifty. And this distemper I take to be the most obstinate, intractable, and disheartening one that can afflict the human machine; and is chiefly produced by intemperate cookery, with its necessary attendant, habitual luxury.

"6. *Gout.*—A total milk and seed diet, with gentle vomits before and after the fits, chewing bark in the morning and rhubarb at night, with bleeding about the equinoxes, will perfectly cure the gout in persons under fifty, and greatly relieve those farther advanced in life; but must be continued ever after, if such desire to get well.

"7. *Gravel.*—Soap lees, softened with a little oil of sweet almonds, drunk about a quarter of an ounce twice a day on a fasting stomach; or soap and egg-shell pills, with a total milk and seed diet, and Bristol water beverage, will either totally dissolve the stone in kidneys or bladder, or render it almost as easy as the nail on one's finger, if the patient is under fifty, and much relieve him, even after that age.

"In about thirty years' practice, in which I have, in some degree or other, advised this method in proper cases, I have had but two patients in whose total recovery I have been mistaken, and these were both scrofulous cases, where the glands and tubercles were so many, so hard, and so impervious that even the ponderous remedies and diet joined could not discuss them; and they were both also too far gone before they entered upon them;—and I have found deep scrofulous vapors the most obstinate of any of this tribe of these distempers. And indeed nothing can possibly reach such, but the ponderous medicines, joined with a liquid, cool, soft, milk and seed regimen; and if these two do not, in due time, I can boldly affirm it, nothing ever will."

Dr. Cheyne goes on to speak of the cure, on similar principles, of a great many other difficult or dangerous diseases, as asthma, pleurisy, hemorrhage, mania, jaundice, bilious colic, rheumatism, scurvy, and venereal disease; but he modestly owns that, in his opinion on these, he does not feel such entire confidence as in the former cases, for want of sufficient experiments. He, however, closes one of his chapters with the following pretty strong statement:

"I am morally certain, and am myself entirely convinced, that a milk and seed, or milk and turnip diet, duly persisted in, with the occasional helps mentioned (elsewhere) on exacerbations, will either totally cure or greatly relieve every chronical distemper I ever saw or read of."

Another chapter is thus concluded, and with it I shall conclude my extracts from his writings.

"Some, perhaps, may controvert, nay, ridicule the doctrine laid down in these propositions. I shall neither reply to, nor be moved with any thing that shall be said against them. If they are of nature and truth, they will stand; if not, I consent they should come to nought. I have satisfied my own conscience—the rest belongs to Providence. Possibly time and bodily sufferings may justify them;—if not to this generation, perhaps to some succeeding one. I myself am convinced, by long and many repeated experience, of their justness and solidity. If what has been advocated through this whole treatise does not convince others, nothing I can add will be sufficient. I will leave only this reflection with my readers.

"All physicians, ancient and modern, allow that a milk and seed diet will totally cure before fifty, and infinitely alleviate after it, the consumption, the rheumatism, the scurvy, the gout—these highest, most mortal, most painful, and most obstinate distempers; and nothing is more certain in mathematics, than that which will cure the greater will certainly cure the lesser distempers."

DR. GEOFFROY.

Dr. Geoffroy, a distinguished French physician and professor of chemistry and medicine in some of the institutions of France, flourished more than a hundred years ago. The bearing of the following extract will be readily seen. It is from the Memoirs of the Royal Academy for the year 1730; and I am indebted for it to the labors of Dr. Cheyne.

"M. Geoffroy has given a method for determining the proportion of nourishment or true matter of the flesh and blood, contained in any sort of food. He took a pound of meat that had been freed from the fat, bones, and cartilages, and boiled it for a determined time in a close vessel, with three pints of water; then, pouring off the liquor, he added the same quantity of water, boiling it again for the same time; and this operation he repeated several times, so that the last liquor appeared, both in smell and taste, to be little different from common water. Then, putting all the liquor together, and filtrating, to separate the too gross particles, he evaporated it over a slow fire, till it was brought to an extract of a pretty moderate consistence.

"This experiment was made upon several sorts of food, the result of which may be seen in the following table. The weights are in ounces, drachms, and grains; sixty grains to a drachm, and eight drachms to an ounce.

Kind of Food.		Amount of Extract.		
		oz.	dr.	gr.
One lb.	Beef	0.	7.	8.
"	Veal	1.	1.	48.
"	Mutton	1.	3.	16.
"	Lamb	1.	1.	39.
"	Chicken	1.	4.	34.
"	Pigeon	1.	0.	12.
"	Pheasant	1.	2.	8.
"	Partridge	1.	4.	34.
"	Calves' Feet	1.	2.	26.
"	Carp	1.	0.	8.
"	Whey	1.	1.	3.
"	Bread	4.	1.	0.

"The relative proportion of the nourishment will be as follows:

Beef	7
Veal	9
Mutton	11
Lamb	9
Chicken	12
Pigeon	8

Pheasant	10
Partridge	12
Calves' Feet	10
Carp	8
Whey	9
Bread	33

"From the foregoing decisive experiments it is evident that white, young, tender animal food, bread, milk, and vegetables are the best and most effectual substances for nutrition, accretion, and sweetening bad juices. They may not give so strong and durable mechanical force, because being easily and readily digestible, and quickly passing all the animal functions, so as to turn into good blood and muscular flesh, they are more transitory, fugitive, and of prompt secretion; yet they will perform all the animal functions more readily and pleasantly, with fewer resistances and less labor, and leave the party to exercise the rational and intellectual operations with pleasure and facility. They will leave Nature to its own original powers, prevent and cure diseases, and lengthen out life."

Now if this experiment proves what Dr. C. supposes in favor of the lighter meats and vegetables taken together, how much more does it prove for bread alone? For it cannot escape the eye of the least observing that this article, though placed last in the list of Dr. Geoffroy, is by far the highest in point of nutriment; nay, that it is about three times as high as any of the rest. I am not disposed to lay so much stress on these experiments as Dr. C. does; nevertheless, they prove something Connected with the more recent experiments of Messrs. Percy and Vauquelin and others, how strikingly do they establish one fact, at least, viz., that bread and the other farinaceous vegetables cannot possibly be wanting in nutriment; and how completely do they annihilate the old-fashioned doctrine—one which is still abroad and very extensively believed—that animal food is a great deal more nourishing than vegetable! No careful inquirer can doubt that bread, peas, beans, rice, etc., are twice as nutritious—to say the least—as flesh or fish.

MESSRS. PERCY AND VAUQUELIN.

As I have alluded, in the preceding article, to the experiments of Messrs. Percy and Vauquelin, two distinguished French chemists, their testimony in this place seems almost indispensable, even though we should not regard it, in the most strict import of the term, as medical testimony. The result of their experiments, as communicated by them to the French minister of the interior, is as follows:

In bread, every one hundred pounds is found to contain eighty pounds of nutritious matter; butcher's meat, averaging the different sorts, contains only thirty-five pounds in one hundred; French beans (in the grain), ninety-two pounds in one hundred; broad beans, eighty-nine pounds; peas, ninety-three pounds; lentils (a species of half pea little known with us), fifty-four pounds in one hundred; greens and turnips only eight pounds of solid nutritious substance in one hundred; carrots, fourteen pounds; and one hundred pounds of potatoes yield only twenty-five pounds of nutriment.

I will just affix to the foregoing one more table. It is inserted in several other works which I have published; but for the benefit of those who may never yet have seen it, and to show how strikingly it corresponds with the results of the experiments of Geoffroy, Percy, and Vauquelin, I deem it proper to insert it.

Of the best wheat, one hundred pounds contain about eighty-five pounds of nutritious matter; of rice, ninety pounds; of rye, eighty; of barley, eighty-three; of beans, eighty-nine to ninety-two; peas, ninety-three; lentils, ninety-four; meat (average), thirty-five; potatoes, twenty-five; beets, fourteen; carrots, ten; cabbage, seven; greens, six; and turnips, four.

DR. PEMBERTON.

Dr. Pemberton, after speaking of the general tendency, in our highly fed communities, to scrofula and consumption, makes the following remarks, which need no comment:

"If a child is born of scrofulous parents, I would strongly recommend that it be entirely nourished from the breast of a healthy nurse, for at least a year. After this, the food should consist of milk and farinaceous vegetables. By a perseverance in this diet for three years, I have imagined that the threatened scrofulous appearances have certainly been postponed, if not altogether prevented."

SIR JOHN SINCLAIR.

Sir John Sinclair, an eminent British surgeon, says, "I have wandered a good deal about the world, my health has been tried in all ways, and, by the aid of temperance and hard work, I have worn out two armies in two wars, and probably could wear out another before my period of old age arrives. I eat no animal food, drink no wine or malt liquor, or spirits of any kind; I wear no flannel; and neither regard wind nor rain, heat nor cold, when business is in the way."

DR. JAMES, OF WISCONSIN.

Dr. James, of Wisconsin, but formerly of Albany, and editor of a temperance paper in that city, one of the most sensible, intelligent, and refined of men, and one of the first in his profession, is a vegetable eater, and a man of great simplicity in all his physical, intellectual, and moral habits. I do not know that his views have ever been presented to the public, but I state them with much confidence, from a source in which I place the most implicit reliance.

DR. CRANSTOUN.

Dr. Cranstoun, a worthy medical gentleman in England, became subject, by some means or other, to a chronic dysentery, on which he exhausted, as it were, the whole materia medica, in vain. At length, after suffering greatly for four or five years, he was completely cured by a milk and vegetable diet. The following is his own brief account of his cure, in a letter to Dr. Cheyne:

"I resolutely, as soon as capable of a diet, held myself close to your rules of bland vegetable food and elementary drink, and, without any other medicine, save frequent chewing of rhubarb and a little bark, I passed last winter and this summer without a relapse of the dysentery; and, though by a very slow advance, I find now more restitution of the body and regularity in the economy, on this primitive aliment, than ever I knew from the beginning of this trouble. This encourages much my perseverance in the same method, and that so religiously, as, to my knowledge, now for more than a year and a half I have not tasted of any thing that had animal life. There is plenty in the vegetable kingdom."

DR. TAYLOR, OF ENGLAND.

This gentleman, who had studied the works of Dr. Sydenham, and was therefore rather favorably inclined toward a milk and vegetable diet, became at last subject to epileptic fits. Not being willing, however, to give up his high living and his strong drinks, he tried the effects of medicine, and even consulted all the most eminent of his brethren of the medical profession in and about London; but all to no purpose, and the fits continued to recur. He used frequently to be attacked with them while riding along the road, in pursuance of the business of his profession. In these cases he would fall from his horse, and often remain senseless till some passenger or wagon came along and carried him to the nearest house. At length his danger, not only from accidents, but from the frequency and violence of the attacks, became so imminent that he was obliged to follow the advice of his master, Sydenham. He first laid aside the use of all fermented and distilled liquors; then, finding his fits became less frequent and violent, he gave up all flesh meat, and confined himself entirely to cows' milk.

In pursuance of this plan, in a year or two the epilepsy entirely left him. "And now," says Dr. Cheyne, from whom I take the account, "for seventeen years he has enjoyed as good health as human nature is capable of, except that once, in a damp air and foggy weather in riding through Essex, he was seized with an ague, which he got over by chewing the bark." He assured Dr. C. that at this time—and he was considerably advanced in life—he could play six hours at cricket without fatigue or distress, and was more active and clear in his faculties than ever he had been before in his whole life. He also said he had cured a great many persons, by means of the same diet, of inveterate distempers.

DRS. HUFELAND AND ABERNETHY.

The celebrated Dr. Hufeland taught that a simple vegetable diet was most conducive to health and long life. The distinguished Dr. Abernethy has expressed an opinion not very unlike it, in the following eccentric manner:

"If you put improper food into the stomach it becomes disordered, and the whole system is affected. Vegetable matter ferments and becomes gaseous, while *animal* substances are changed into a putrid, abominable, and acrid stimulus. Now, some people acquire preposterous noses; others, blotches on the face and different parts of the body; others, inflammation of the eyes; all arising from the irritations of the stomach. I am often asked why I don't practice what I preach. I reply by reminding the inquirer of the parson and sign-post—both point the way, but neither follows its course."

DR. GREGORY.

Dr. Gregory, a distinguished professor and practitioner of medicine in Scotland, in a work published more than seventy years ago, strongly recommends plain and simple food for children. Till they are three years old, he says, their diet should consist of plain milk, panada, good bread, barley meal porridge, and rice. He also complains of pampering them with animal food. The same arguments which are good for forming them to the habits of vegetable food exclusively for the first three years of life, would be equally good for its continuance.

DR. CULLEN, OF EDINBURGH.

The name of Dr. Cullen is well known, and he has long been regarded as high authority. Yet this distinguished writer and teacher expressly says, that a very temperate and *sparing* use of animal food is the surest means of preserving health and obtaining long life. But I will quote his own language, in various parts of his writings. And first, from his Materia Medica:

"Vegetable aliment, as never over-distending the vessels or loading the system, never interrupts the stronger emotions of the mind, while the heat, fullness, and weight of animal food, is an enemy to its vigorous efforts. Temperance, then, does not consist so much in the quantity, for that will always be regulated by our appetite, as in the *quality*, viz., a large proportion of vegetable aliment."

I will not stop here to oppose Dr. C.'s views in regard to the quantity of our food; for this is not the place. It is sufficient to show that he admits the importance of *quality*, and gives the preference to a diet of vegetables.

He seems in favor, in another place in his works, of sleeping after eating—perhaps a heresy, too—and inclines to the opinion that the practice would be hardly hurtful if we ate less animal food.

But his "First Lines of the Practice of Physic," abounds in testimonies in favor of vegetable food. In speaking, for example, of the cure of rheumatic affections, he has the following language:

"The cure, therefore, requires, in the first place, an antiphlogistic regimen, and particularly, a total abstinence from animal food, and from all fermented or spirituous liquors."

"Antiphlogistic regimen," in medical language, means that food and drink which is most cooling and quieting to the stomach and to the general system.

In the treatment of gout, Dr. Cullen recommends a course like that which has been stated, except that instead of proposing vegetable food as a means of cure, he recommends it as *preventive*. He says—

"The gout may be entirely prevented by constant bodily exercise, and by a low diet; and I am of opinion that this prevention may take place even in persons who have a hereditary disposition to the disease. I must add, here, that even when the disposition has discovered itself by severe paroxysms of inflammatory gout, I am persuaded that labor and abstinence will absolutely prevent any returns of it for the rest of life."

Again, in reference to the same subject, he thus observes:

"I am firmly persuaded that any man who, early in life, will enter upon the constant practice of bodily labor and of abstinence from animal food, will be preserved entirely from the disease."

And yet once more.

"If an abstinence from animal food be entered upon early in life, while the vigor of the system is yet entire, I have no doubt of its being both safe and effectual."

To guard against the common opinion that by vegetable food, he meant raw, or crude, or bad vegetables, Dr. C. explains his meaning by assuring the reader that by a vegetable diet he means the "farinaceous seeds," and "milk;" and admits that green, crude, and bad vegetables are not only less useful, but actually liable to produce the very diseases, which good, mealy vegetable food will prevent or cure.

This is an important distinction. Many a person, who wishes to be abstemious, seems to think that if he only abstains from flesh and fish, that is enough. No matter, he supposes, what vegetables he uses, so they are vegetables; nor how much he abuses himself by excess in quantity. Nay, he will even load his stomach with milk, or butter, or eggs; sometimes with fish (we have often been asked if we considered fish as animal food); and sometimes, worse still, with hot bread, hot buckwheat cakes, hot short-cakes, swimming, almost, in butter;—yes, and sometimes he will even cover his potatoes with gravy, mustard, salt, etc.

It is in vain for mankind to abstain from animal food, as they call it, and yet run into these worse errors. The lean parts of animals not much fattened, and only rarely cooked, eaten once a day in small quantity, are far less unwholesome than many of the foregoing.

But to return to Dr. C. In speaking of the proper drink for persons inclined to gout, he thus remarks:

"With respect to drink, fermented liquors are useful only when they are joined with animal food, and that by their acescency; and their stimulus is only necessary from custom. When, therefore, animal food is to be avoided, fermented liquors are unnecessary, and by increasing the acescency of vegetables, these liquors may be hurtful. The stimulus of fermented or spirituous liquors is not necessary to the young and vigorous: and, when much employed, impairs the tone of the system."

Dr. C. might have added—what indeed we should infer by parity of reasoning—that when fermented liquors are avoided, animal food is no longer necessary, and by increasing the alkaline state of the stomach and fluids, may be hurtful. The truth is, they go best together. If we use flesh and fish, which are alkaline, a small quantity of gently acid drink, as weak cider or wine, taken either *with* our meals, or *between* them, may be useful. It is better, however, to abstain from both.

For if a purely vegetable aliment, with water alone for drink, is safe to all young persons inclining at all to gout, to whom is it unsafe? If it tends to render a young person at all weaker, that very weakness would predispose to the gout, in some of its forms, if a person were constitutionally inclined to that disease—if not to some other complaint, to which he was more inclined. It cannot, therefore, be unsafe to any, if Dr. C. is right.

But if those who are trained to it, *lose* nothing, even in the high latitude of Scotland—where Dr. C. wrote—by confining themselves to good vegetables and water, then they must necessarily *gain*, on his own principles, by this way of living, because they get rid of any sort of necessity (he might have added, lose their appetite) for fermented liquors.

More than this, as the doctor himself concludes, in another place, they prevent many acute diseases. His words are these:—"It is animal food which especially predisposes to the plethoric and inflammatory state; and that food is therefore to be especially avoided." It is true, he is here speaking of gouty persons: but his principles are also fairly susceptible, as I have shown, of a general application.

In short, it is an undeniable fact, that even a thorough-going vegetable eater might prove every thing he wished, from old established writers on medicine and health, though themselves were feeders on animal food; just as a teetotaler may prove the doctrine of abstinence from all drinks but water, from the writings of medical men, though themselves are still, in many cases, pouring down their cider, their beer, or their wine—or at least, their tea and coffee.

DR. BENJAMIN RUSH.

I find nothing in the writings of this great man which shows, with certainty, what his views were, in regard to animal food. The presumption is, that he was sparing in its use, and that he encouraged a very limited use of it in others. This is presumed, 1, from the general tenor of his writings—deeply imbued as they are with the great doctrine of temperance in all things; and, 2, from the fondness he seems to have manifested in mentioning the temperance and even abstinence of individuals of whom he was speaking.

Of Ann Woods, for example, who died at the age of ninety-six years, he says, "Her diet was simple, consisting chiefly of weak tea, milk, cheese, butter, and vegetables. Meat of all kinds, except veal, disagreed with her stomach. She found great benefit from frequently changing her aliment. Her drinks were water, cider and water, and molasses and vinegar in water. She never used spirits. Her memory (at her death) was but little impaired. She was cheerful, and thankful that her condition in life was happier than that of hundreds of other people."

In his account of Benjamin Lay, a philosopher of the sect of the Friends, in Pennsylvania, Dr. R. relates, that "he was extremely temperate in his diet, living chiefly upon vegetables. Turnips boiled and afterward roasted, were his favorite dinner. His drink was pure water. He lived above eighty years." It appears, also, that he was exceedingly healthy.

He relates of Anthony Benezet, a distinguished teacher of Philadelphia, who lived to an advanced age, that his sympathy was so great with every thing that was capable of feeling pain, that he resolved, toward the close of his life, to eat no animal food. He also relates the following singular anecdote of him. Upon coming into his brother's house, one day, when the family were dining upon poultry, he was asked by his brother's wife to sit down and dine with them. What! said he, would you have me eat my neighbors?

Dr. Caleb Bannister, in another part of this work, tells us that he was led to adopt a milk and vegetable diet, in incipient consumption, from reading the writings of Dr. Rush; and I have little doubt that Dr. R. himself lived quite abstemiously, if not altogether on vegetables.

Nor is this *incidental* testimony from Dr. Rush quite all. In his work "On the Diseases of the Mind," he speaks often of the evils of eating high-seasoned food, and especially animal food. And in stating what were the proper remedies for debility in young men, when induced by certain forms of licentiousness, he expressly insists on a diet consisting simply of vegetables, and prepared without condiments; and he even encourages the disuse of salt. Had Dr. Rush lived to this day, he would, ere now, in all probability, have fully adopted and defended the vegetable system. With views like his on the subject of intemperance, and a mind ever open to conviction, the result could hardly have been otherwise.

DR. WILLIAM LAMBE, OF LONDON.

Dr. William Lambe, of London, is distinguished both as a physician and a general scholar, and is a prominent member of the "College of Physicians." He was a graduate of St. John's College, Cambridge, and a fellow-student with the immortal Clarkson.

Dr. Lambe is the author of several valuable works, among which are his "Reports on Cancer," and a more recent work entitled, "Additional Reports on the Effects of a Peculiar Regimen, in Cases of Cancer, Scrofula, Consumption, Asthma, and other chronic diseases." He has also made and published numerous experiments, especially in chemistry, which is, with him, a favorite science; and it is said that he has spent fortunes in this way.

Dr. L. is now eighty-four years of age, and has lived on vegetable diet forty-two years. He commenced this course to cure himself of internal gout, and continued it because he found it better for his health. He is now only troubled with it slightly, at his extremities, which he thinks highly creditable to a vegetable course—having thrown it off from his vital organs. He is cheerful and active, and able to discharge the duties of an extensive medical practice. He walks into town, a distance of three miles from his residence, every morning, and back at night; and thinks himself as likely to live twenty years longer as he was, twenty years ago, to live to his present age.

The following is a condensed account of Dr. L.'s views, as obtained from his "Additional Reports," above mentioned. Some of the first paragraphs relate to the effects of vegetable food on those who are predisposed to scrofula, consumption, etc.

"We see daily examples of young persons becoming consumptive who never went without animal food a single day of their lives. If the use of animal food were necessary to prevent consumption, we should expect, where people lived almost entirely upon such a diet, the disease would be unknown.

"Now, the Indian tribes visited by Mr. Hearne live in this manner. They do not cultivate the earth. They subsist by hunting, and the scanty produce of spontaneous vegetation. But, among these tribes consumption is common. Their diseases, as Mr. Hearne informs us, are principally fluxes, scurvy, and consumption.

"In the last four years, several cases of glandular swellings have occurred to me at the general dispensary, and I have made particular inquiries into the mode of living of such children. In the majority, they had animal food. In opposition to the accusation of vegetable food causing tumefaction of the abdomen, I must testify, that twice in my own family I have seen such swellings disappear under a vegetable regimen, which had been formed under a diet of animal food.

"Increasing the strength, for a time, is no proof of the salubrity of diet. The increased strength may not continue, though the diet should be continued. On the contrary, there is a sort of oscillation; the strength just rising, then sinking again. This is what is experienced by the trainers of boxers. A certain time is necessary to get these men into condition; but this condition cannot be maintained for many weeks together, though the process by which it was formed is continued. The same is found to hold in the training of race-horses, and fighting-cocks.

"It seems certain that animal food predisposes to disease. Timoric, in his account of the plague at Constantinople, asserts that the Armenians, who live chiefly on vegetable food, were far less disposed to the disease than other people. The typhus fever is greatly exasperated by full living.

"It seems, moreover, highly probable that the power inherent in the human living body, of restoring itself under accidents or wounds, is strongest in those who use most a vegetable regimen.

"Contagions act with greater virulence upon bodies prepared by a full diet of animal food.

"Since fishing has declined in the isles of Ferro, and the inhabitants have lived chiefly on vegetables, the elephantiasis has ceased among them.

"Those monks who, by the rules of their institution, abstain from the flesh of animals, enjoy a longer mean term of life, as the consequence. Of this there can be no doubt. Of one hundred and fifty-two monks, taken promiscuously in all times and all sorts of climates, there lives produced a total, according to Baillot (a writer of eminence), of 11,589 years, or an average of seventy-six years and a little more than three months.

"Those Bramins who abstain most scrupulously from the flesh of animals attain to the greatest longevity.

"Life is prolonged, under incurable diseases, about one tenth by vegetable diet; so that a person who would otherwise die at seventy, will reach seventy-seven. In general, however, the proportion is about one sixth.

"Abstaining from animal food palliates, when it does not cure, all constitutional diseases.

"The use of animal food hurries on life with an unnatural and unhealthy rapidity. We arrive at puberty too soon; the passions are developed too early; in the male, they acquire an impetuosity approaching to madness; females become mothers too early, and too frequently; and, finally, the system becomes prematurely exhausted and destroyed, and we become diseased and old, when we ought to be in middle life.

"It affords no trifling ground of suspicion against the use of animal food that it so obviously inclines us to corpulency. Corpulency itself is a species of disease, and a still surer harbinger of other diseases. It is so even in animals. When a sheep has become fat, the butcher knows it must be killed or it will rot and decline. It is rare indeed for the corpulent to be long-lived. They are at the same time sleepy, lethargic, and short-breathed. Even Hippocrates says, 'Those who are uncommonly fat die more quickly than the lean.'

"As a general, rule, the florid are less healthy than those who have little color; an increase of color having ever been judged, by common sense, to be a sign of impending illness. Some, however, who are lean upon animal food, thrive upon vegetables, and improve in color.

"All the notions of vegetable diet affording only a deficient nutriment—notions which are countenanced by the language of Cullen and other great physicians—are wholly groundless.

"Man is herbivorous in his structure.

"I have observed no ill consequences from the relinquishment of animal food. The apprehended danger of the change, with which men scare themselves and their neighbors, is a mere phantom of the imagination. The danger, in truth, lies wholly on the other side.

"There is no organ of the body which, under the use of vegetable food, does not receive an increase of sensibility, or of that power which is thought to be imparted to it by the nervous system.

"Socrates, Plato, Zeno, Epicurus, and others of the masters of ancient wisdom, adhered to the Pythagorean diet (vegetable diet), and are known to have arrived at old age with the enjoyment of uninterrupted health. Celsus affirms that the bodies which are filled with much animal food become the most quickly old and diseased. It was proverbial that the ancient athletæ were the most stupid of men. The cynic Diogenes, being asked what was the cause of this stupidity, is reported to have answered, 'Because they are wholly formed of the flesh of swine and oxen.' Theophrastus says that feeding upon flesh destroys the reason, and makes the mind more dull.

"Animal food is unfavorable to the intellectual powers. The effect is, in some measure, instantaneous; it being hardly possible to apply to any thing requiring thought after a full meal of meat; so that it has been not improperly said of vegetable feeders, that *with them it is morning all day long.* But the senses, the memory, the understanding, and the imagination have also been observed to improve by a vegetable diet.

"It will not be disputed that, for consumptive symptoms, a vegetable diet, or at least a vegetable and milk diet, is the most proper.

"It has been said, that the great fondness men have for animal food, is proof enough that nature intended them to eat it. As if men were not fond of wine, ardent spirits, and other things which we know cut short their days!

"In every period of history it has been known that vegetables alone are sufficient for the support of life; and the bulk of mankind live upon them at this hour. The adherence to the use of animal food is no more than a gross persistence in the customs of savage life, and an insensibility to the progress of reason and the operation of intellectual improvement. This habit must be considered as one of the numerous relics of that ancient barbarism which has overspread the face of the globe, and which still taints the manners of civilized nations.

"The use of fermented liquors is, in some measure, a necessary concomitant and appendage to the use of animal food. Animal food, in a great number of persons, loads the stomach, causes some degree of oppression, fullness, and uneasiness; and, if the measure of it be in excess, some nausea and tendency to sickness. Such persons say meat is too heavy for the stomach. Fish is still more apt to nauseate. The use of fermented liquors takes off these uneasy feelings, and is thought to assist digestion. In short, in the use of animal food, man having deviated from the simple aliment offered him by the hand of nature, and which is the best suited to his organs of digestion, he has brought upon himself a premature decay, and much intermediate suffering connected with it. To this use of animal food almost all nations that have emerged from a state of barbarism, have united the use of spirituous and fermented liquors."

It is but justice to Dr. L., however, as the above was written by him over thirty years ago, to say, that though he still adheres to the same views, he thinks pure distilled water a very important addition to the vegetable diet, in the cure of chronic diseases. The following are his remarks in a letter to Mr. Graham, dated ten or twelve years ago.

"My doctrine is, that for the preservation of health, and more particularly for the successful treatment of chronic diseases, it is necessary to attend to the *whole* ingesta—to the *fluid* with as much care as the solid. And I am persuaded that the errors into which men have fallen with regard to supposed mischiefs or inconveniences (as weakness, for example), as resulting from a restriction to a vegetable diet, have, to a very considerable extent arisen from a want of a proper attention to the quality of the water they drank. So far back as the year 1803, I found that the use of pure distilled, instead of common water, relieved a state of habitual suffering of the stomach and bowels. On this account, I always require that *distilled* water shall be joined to the use of a vegetable diet; and consider this to be essential to the treatment."

PROFESSOR LAWRENCE.

Professor Lawrence is the author of a work entitled Lectures on Physiology, Zoology, and the Natural History of Man. He is a member of the Royal College of Surgeons, London, Professor of Anatomy and Surgery to the College, and Surgeon to several Hospitals. In his work above mentioned, after much discussion in regard to the natural dietetic character of man, he thus remarks:

"That animal food renders man strong and courageous, is fully disproved by the inhabitants of northern Europe and Asia, the Laplanders, Samoiedes, Ostiacs, Tungooses, Burats, and Kamtschadales, as well as by the Esquimaux in the northern, and the natives of Terra del Fuego in the southern extremity of America, which are the smallest, weakest, and least brave people of the globe, although they live almost entirely upon flesh, and that often raw.

"Vegetable diet is as little connected with weakness and cowardice, as that of animal matter is with physical force and courage. *That men can be perfectly nourished, and their bodily and mental capabilities fully developed in any climate, by a diet purely vegetable, admits of abundant proof from experience.* In the periods of their greatest simplicity, manliness, and bravery, the Greeks and Romans appear to have lived almost entirely on plain vegetable preparations. Indifferent bread, fruits, and other produce of the earth, are the chief nourishment of the modern Italians, and of the mass of the population in most countries in Europe. Of those more immediately known to ourselves, the Irish and Scotch may be mentioned, who are certainly not rendered weaker than their English fellow-subjects by their free use of vegetable aliment. The Negroes, whose great bodily powers are well known, feed chiefly on vegetable substances; and the same is the case with the South Sea Islanders, whose agility and strength were so great that the stoutest and most expert English sailors had no chance with them in wrestling and boxing."

The concession of Prof. L., which I have placed in italic, is sufficient for our purpose; we ask no more. Nevertheless, I am willing to hear his views of the indications afforded by our anatomical character, which are, as will be seen, equally decisive in favor of vegetable eating.

"Physiologists have usually represented that our species holds a middle rank, in the masticatory and digestive apparatus, between the flesh-eating and herbivorous animals—a statement which seems rather to have been deduced from what we have learned by experience on the subject, than to result from an actual comparison of men and animals.

"The teeth and jaws of men are, in all respects, much more similar to those of monkeys than of any other animal. The number is the same as in man, and the form so closely similar, that they might easily be mistaken for human. In most of them, except the ourang-outang, the canine teeth are much larger and stronger than in us; and so far, these animals have a more carnivorous character than man.

"Thus we find, that whether we consider the teeth and jaws, or the immediate instruments of digestion, the human structure closely resembles that of the simiæ (monkey race), all of which, in their natural state, are completely herbivorous. Man possesses a tolerably large cœcum, and a cellular colon; which I believe are not found in any herbivorous animal."

The ourang-outang naturally prefers fruits and nuts, as the professor himself shows by extracts from the statements of travelers and naturalists. He is also fond of bread. On board a ship or elsewhere, *in confinement*, he may, however, be taught, like men, to eat almost any thing;—not only to eat milk and suck eggs, but even to eat raw flesh.

It is true, indeed, after all these foregoing statements and concessions in regard to man's native character and the wholesomeness of a diet exclusively vegetable—and after admitting that the human body and mind can be fully and perfectly nourished and *developed* on it, this distinguished writer goes on to say that it is still doubtful which diet—animal, vegetable, or mixed—is on the whole *most* conducive to health, and strength—which is best calculated to avert or remove disease—whether errors in quantity or quality are most pernicious, etc. He says the solution of these and other analogous questions, can only be expected from experimental investigation. He proceeds to say—

"*Mankind are so averse to relinquish their favorite indulgences, and to desert established habits*, that we cannot entertain very sanguine expectations of any important discovery in this department. We must add to this, that there are many other causes affecting human health, besides diet. Before venturing to draw any inferences on a subject beset with so many obstacles, it would be necessary to observe the effects of a purely animal and a purely vegetable diet on several individuals of different habits, pursuits, and modes of life; to note their state, both bodily and mental; and to learn the condition of two or three generations fed in the same manner."

Now, the only difference between this opinion and what I conceive to be the truth in the case is, that just such experimental investigations as those to which he refers have, to all intents and purposes, been already made; as, I trust, will be distinctly shown in the sequel of this work.

DR. SALGUES.

Dr. Salgues, Physician, and Professor of Anatomy, Physiology, etc., etc., to the Institute of France, some years ago wrote a book, entitled "Rules for Preserving the Health of the Aged," which contained many very judicious remarks on diet. There is nothing in the volume, however, which is decidedly in favor of a diet exclusively vegetable, unless it is a few anecdotes; and I have introduced his name chiefly as a sort of authority for those anecdotes. They are the following:

"Josephus informs us that the Essenes were very long lived; many lived upward of one hundred years, solely from their simple habits and sobriety. Aristotle and Plato speak of Herodicus the philosopher, who, although of a feeble and consumptive habit, lived, in consequence of his sobriety, upward of one hundred years. Phabrinus, mentioned by Athenius, lived more than one hundred years, drinking milk only. Zoroaster, according to Pliny, remained twenty years in a desert, living on a small quantity of cheese only."

THE AUTHOR OF "SURE METHODS," ETC.

The British author of "Sure Methods of Improving Health and Prolonging Life," supposed by many to be the distinguished Dr. Johnson, speaks thus:

"It must be confessed that, in temperate climates, at least, an animal diet is, in one respect, more wasting than a vegetable, because it excites, by its stimulating qualities, a temporary fever after every meal, by which the springs of life are urged into constant, preternatural, and weakening exertions. Again;

persons who live chiefly on animal food are subject to various acute and fatal disorders, as the scurvy, malignant ulcers, inflammatory fevers, etc., and are likewise liable to corpulency, more especially when united to inordinate quantities of liquid aliment. There appears to be also a tendency in an animal diet to promote the formation of many chronic diseases; and we seldom find those who indulge much in this diet to be remarkable for longevity.

"In favor of vegetables, it may be justly said, that man could hardly live entirely on animal food, but we know he may on vegetable. Vegetable aliment has likewise no tendency to produce those constitutional disorders which animal food so frequently occasions. And this is a great advantage, more especially in our country (he means in Great Britain), where the general sedentary mode of living so powerfully contributes to the formation and establishment of numerous severe chronic maladies. Any unfavorable effects vegetable food may have on the body, are almost wholly confined to the stomach and bowels, and rarely injure the system at large. This food has also a beneficial influence on the powers of the mind, and tends to preserve a delicacy of feeling, and liveliness of imagination, and acuteness of judgment, seldom enjoyed by those who live principally on meat. It should also be added, that a vegetable diet, when it consists of articles easily digested, as potatoes, turnips, bread, biscuit, oatmeal, etc., is certainly favorable to long life."

BARON CUVIER.[10]

Perhaps it is not generally known that Baron Cuvier, the prince of naturalists, in the progress of his researches came to the most decisive conclusion, that, so far as any thing can be ascertained or proved by the investigation of science in regard to the natural dietetic character of man, he is a fruit and vegetable eater. I have not seen his own views; but the following are said, by an intelligent writer, to be a tolerably faithful transcript of them, and to be derived from his Comparative Anatomy.

"Man resembles no carnivorous animal. There is no exception, unless man be one, to the rule of herbivorous animals having cellulated colons.

"The ourang-outang perfectly resembles man, both in the order and number of his teeth. The ourang-outang is the most anthropomorphous of the ape tribe, all of which are strictly frugivorous. There is no other species of animals, which live on different food, in which this analogy exists. In many frugivorous animals, the canine teeth are more pointed and distinct than those of man. The resemblance also of the human stomach to that of the ourang-outang, is greater than to that of any other animal.

"The intestines are also identical with those of herbivorous animals, which present a large surface for absorption, and have ample and cellulated colons.

The cœcum also, though short, is larger than that of carnivorous animals; and even here the ourang-outang retains its accustomed similarity.

"The structure of the human frame, then, is that of one fitted to a pure vegetable diet, in every essential particular. It is true, that the reluctance to abstain from animal food, in those who have been long accustomed to its stimulus, is so great in some persons of weak minds, as to be scarcely overcome; but this is far from being any argument in its favor. A lamb, which was fed for some time on flesh by a ship's crew, refused its natural diet at the end of the voyage. There are numerous instances of horses, sheep, oxen, and even wood-pigeons, having been taught to live upon flesh, until they have loathed their natural aliment."

No one will deny that Baron Cuvier was in favor of flesh eating; but it was not because he ever believed, for one moment, that man was *naturally* a flesh-eating animal. Man is a reasoning animal (he argues), and intended to be so. If left to the guidance of his instincts, the same yielding to the law of his structure which would exclude flesh meats, should also exclude cookery. Or, in other words, if he is not permitted to depart from the line of life which his structure indicates, he must no more cook his vegetables than eat animal food. Besides, he is made, as Cuvier supposes, for artificial society, and the Creator designed him to *improve* his food; and, if I understand his reasoning, he is better able, with his present structure of teeth, jaws, stomach, intestines, etc., to make this improvement, and rise above his nature, and yield to the force and indications of reason and experience, than if he possessed any other known living structure.

To this structure, however, as well as to the same power of adaptation, the monkey race, and especially the ourang-outang, closely typo approximates. Cuvier's reasoning, in my view, applies only to the adaptability (if I may be allowed the expression) of the human animal, without deciding how far he should avail himself of his power to make changes.

DR. LUTHER V. BELL.

I have alluded, in another part of this work, to the prize essay of Dr. Bell, awarded to him by the Boylston Medical Committee on the subject of the diet of laborers in New England. Dr. Bell is a physician of respectable talents, and is at present the Physician to an Insane Hospital in Charlestown, near this city.

Dr. Bell admits, with the most distinguished naturalists and physiologists of Europe,—Cuvier, Lawrence, Blumenbach, Bell of London, Richerand, Marc, etc.,—that the structure of man resembles closely that of the monkey race; and hence objects to the conclusion to which some of these men have arrived (by jumping over, as it were), that man is an omnivorous animal. He freely allows—I use his own words—"that man does approximate more closely to the frugivorous animals than to any others, in physical organization." But then he insists that the conclusion which ought to be drawn from this similarity "is, that he is designed to have his food in about the same state of mechanical cohesion, requiring about the same energy of masticatory organs, as if it consisted of fruits, etc., alone."

But, wherefore should we draw even this conclusion, if structure and instinct prove nothing, and if we are to be governed solely by reason, without regard to structure and instinct? For my own part, I believe reason is never true reason, when it turns wholly out of doors either instinct or the indications of organization. In other words, an enlightened reason would look both to the structure and organization of man, and to a large and broad experience, for the solution of a question so important as what diet is, on the whole, best for man. And the experience of the world, both in the present and all former ages, leads me to a conclusion entirely different from that to which Dr. Bell, and those who entertain the same views with him, seem to have arrived—a conclusion which is indicated by structure, and confirmed by facts and universal experience. But this subject will be further discussed and developed in another place. It is sufficient for my present purpose, to bring testimony in favor of the safety of vegetable eating, and of the doctrine that man is naturally a vegetable and fruit-eating animal; and especially if I produce, to this end, the testimony of flesh-eaters themselves.

DR. WILLIAM BUCHAN, AUTHOR OF "DOMESTIC MEDICINE."

"Indulgence in animal food, renders men dull and unfit for the pursuits of science, especially when it is accompanied with the free use of strong liquors. I am inclined to think that *consumptions*, so common in England, are, in part, owing to the great use of animal food. But the disease most common to this country is the scurvy. One finds a dash of it in almost every family, and in some the taint is very deep. A disease so general must have a general cause, and there is none so obvious as the great quantity of animal food which is devoured. As a proof that scurvy arises from this cause, we are in possession of no remedy for that disease equal to the free use of fresh vegetables. By the uninterrupted use of animal food, a putrid diathesis is induced in the system, which predisposes to a variety of disorders. I am fully convinced that many of those obstinate complaints for which we are at a loss to account, and which we find it still more difficult to cure, are the effects of a scorbutic taint, lurking in the habit.

"The choleric disposition of the English is almost proverbial. Were I to assign a cause, it would be, their living so much on animal food. There is no doubt but this induces a ferocity of temper unknown to men whose food is taken chiefly from the vegetable kingdom.[11]

"Experience proves that not a few of the diseases incident to the inhabitants of this country, are owing to their mode of living. The vegetable productions they consume, fall considerably short of the proportion they ought to bear to the animal part of their food. The major part of the aliment ought to consist of vegetable substances. There is a continual tendency in animal food, as well as in the human body itself, to putrefaction; which can only be counteracted by the free use of vegetables. All who value health, ought to be contented with making one meal of animal food in twenty-four hours; and this ought to consist of one kind only.

"The most obstinate scurvy has often been cured by a vegetable diet; nay, milk alone, will frequently do more in that disease than any medicine. Hence it is evident that if vegetables and milk were more used in diet, we should have less scurvy, and likewise fewer putrid and inflammatory fevers.

"Such as abound with blood (and such are almost all of us), should be sparing in the use of every thing which is highly nourishing—as fat meat, rich wines, strong ales, and the like. Their food should consist chiefly of bread and other vegetable substances; and their drink ought to be water, whey, or small beer."

Dr. B. also insists on a vegetable diet, as a preventive of many diseases; particularly of consumption. When there is a tendency to this disease, in the young, he says "it should be counteracted by strictly adhering to a diet of the farinacea, and ripe fruits. Animal food and fermented liquors ought to be rigidly prohibited. Even milk often proves too nutritious."

DR. CHARLES WHITLAW.

Dr. Whitlaw is the author of a work entitled "New Medical Discoveries," in two volumes, and of a "Treatise on Fever." He has also established medical vapor baths in London, New York, and elsewhere; and is a gentleman of much skill and eminence in his profession. Dr. Whitlaw says—

"All philosophers have given their testimony in favor of vegetable food, from Pythagoras to Franklin. Its beneficial influence on the powers of the mind has been experienced by all sedentary and literary men.

"But, that which ought to convince every one of the salubrity of a diet consisting of vegetables, is the consideration of the dreadful effects of totally abstaining from it, unless it be for a very short time; accounts of which we meet with, fully and faithfully recorded, in the most interesting and most authentic narratives of human affairs—wars, sieges of places, long encampments, distant voyages, the peopling of uncultivated and maritime countries, remarkable pestilences, and the lives of illustrious men. To this cause the memorable plague at Athens was attributed; and indeed all the other plagues and epidemical distempers, of which we have any faithful accounts, will be found to have originated in a deprivation of vegetable food.

"The only objections I have ever heard urged (the only plausible ones, he must mean, I think), is the notion of its inadequacy to the sustenance of the body. But this is merely a strong prejudice into which the generality of mankind have fallen, owing to their ignorance of the laws of life and health. Agility and constant vigor of body are the effect of health, which is much better preserved by a herbaceous, aqueous, and sparing tender diet, than by one which is fleshy, vinous, unctuous, and hard of digestion.

"So fully were the Romans, at one time, persuaded of the superior goodness of vegetable diet, that, besides the private example of many of their great men, they established laws respecting food, among which were the *lex fannia*, and the *lex licinia*, which allowed but very little animal food; and, for a period of five hundred years, diseases were banished along with the physician from the Roman empire. Nor has our own age been destitute of examples of men, brave from the vigor both of their bodies and their minds, who at the same time have been drinkers of water and eaters of vegetables.[12]

"Nothing is more certain than that animal food is inimical to health. This is evident from its stimulating qualities producing, as it were, a temporary fever after every meal; and not only so, but from its corruptible qualities it gives rise to many fatal diseases; and those who indulge in its use seldom arrive at an advanced age.

"We have the authority of the Scripture for asserting that the proper aliment of man is vegetables. See Genesis. And as disease is not mentioned as a part of the cause, we have reason to believe that the antediluvians were strangers to this evil. Such a phenomenon as disease could hardly exist among a people who lived entirely on a vegetable food; consequently all the individuals made mention of in that period of the world, are said to have died of old age; whereas, since the day of Noah, when mankind were permitted to eat animal food, such an occurrence as a man dying of old age, or a natural decay of the bodily functions, does not occur probably once in half a century.

"Its injurious effects on the mind are equally certain. The Tartars, who live principally on animal food, are cruel and ferocious in their disposition, gloomy and sullen minded, delighting in exterminating wars and plunder; while the Bramins and Hindoos, who live entirely on vegetable aliment, possess a mildness and gentleness of character and disposition directly the reverse of the Tartar; and I have no doubt, had India possessed a more popular form of government, and a more enlightened priesthood, her people, with minds so fitted for contemplation, would have far outstripped the other nations of the world in manufactures, and in the arts and sciences.

"But we need only look at the peasantry of Ireland, who, living as they do, chiefly on a vegetable—and to say the least of it, a very suspicious kind of aliment, I mean the potatoe—are yet as robust and vigorous a race of men as inherit any portion of the globe.

"The greater part of our bodily disease is brought on by improper food. This opinion has been strongly confirmed by my daily experience in the treatment of those diseases to which the people of England are peculiarly subject, such as scrofula, consumption, leprosy, etc. These disorders are making fearful and rapid strides; so much so, that not a single family may now be considered exempt from their melancholy ravages."

This is fearful testimony, but it is the result of much observation and of twenty years' experience. But the same causes are producing the same effects—at least, so far as scrofula and consumption are concerned—in this country, at the present time, of which Dr. W. complains so loudly in England. I could add much more from his writings, but what I have said is sufficient.

DR. JAMES CLARK.

Dr. Clark, physician to the king and queen of Belgium, in a Treatise on Pulmonary Consumption, has the following remarks:

"There is no greater evil in the management of children than that of giving them animal diet very early. By persevering in the use of an over-stimulating diet, the digestive organs become irritated, and the various secretions immediately connected with and necessary to digestion are diminished, especially the biliary secretion; and constipation of the bowels and congestion of the abdominal viscera succeed. Children so fed, moreover, become very liable to attacks of fever and of inflammation, affecting particularly the mucous membranes; and measles and the other diseases incident to childhood are generally severe in their attack."

The suggestion that a mild or vegetable diet will render certain diseases incident to childhood more mild than otherwise they would be, is undoubtedly an important one; and as just as it is important. But the remark might be extended, in its application. Both children and adults would escape all sorts of diseases, especially colds and epidemics, with much more certainty, or, if attacked, the attacks would be much more mild, on an exclusively vegetable diet than on a mixed one. Dr. Clark does not, indeed, say so; but I may say it, and with confidence. And Dr. C. could not probably show any reason why, on his own principles, it should not be so.

PROF. MUSSEY, OF DARTMOUTH COLLEGE.

Prof. R. D. Mussey, of Hanover, New Hampshire, whose science and skill as a surgeon and physician are well known and attested all over New England, has for many years taught, both directly and indirectly, in his public lectures, that man is naturally a fruit and vegetable eater. This he proves, first, from the structure of his teeth and intestines—next from his physiological character, and finally, from various facts and considerations too numerous to detail here.

He thinks the Bible doctrines are in favor of the disuse of flesh and fish; that the Jews were required to abstain from pork, and from all fat and blood, for physiological no less than other reasons. An infant, he says, naturally has a disrelish for animal food. He says that, in all probability, animal food was not permitted, though used, before the flood; and that its use, contrary to the wish of the Creator, was probably one cause of human degeneracy. Animal food, he says, is apt to produce diseases of the skin—makes people passionate and violent—excites the nervous system too much—renders the senses and faculties more dull—and favors the accumulation of what is mired tartar on the teeth, and thus causes their early and certain decay. The blood and breath of carnivorous animals emit an unpleasant odor, while those of vegetable eaters do not. The fact that man *does eat* flesh no more proves its necessity, than the fact that cows, and sheep, and horses can be taught it, proves its necessity to them. The Africans bear the cold better the first winter after their arrival in a northern climate than afterward. May not this be owing to their simple vegetable living?

DR. CONDIE, OF PHILADELPHIA.

The Journal of Health, edited by some of the ablest physicians of Philadelphia, has the following remarkable language on the subject of vegetable food. See vol. 1, page 277.

"It is well known that vegetable substances, particularly the farinaceous, are fully sufficient, of themselves, for maintaining a healthy existence. We have every reason for believing that the fruits of the earth constituted, originally, the only food of man. Animal food is digested in a much shorter period than vegetables; from which circumstance, as well as its approaching much nearer in its composition to the substance of the body into which it is to be converted, it might at first be supposed the most appropriate article of nourishment. It has, however, been found that vegetable matter can be as readily and perfectly *assimilated* by the stomach into appropriate *nutriment* as the most tender animal substances; and confessedly with a less heating effect upon the system generally.

"As a general rule, it will be found that those who make use of a diet consisting chiefly of vegetable matter have a vast advantage in looks, in strength, and spirits, over those who partake largely of animal food. They are remarkable for the firm, healthy plumpness of their muscles, and the transparency of their skins. This assertion, though at variance with popular opinion, is amply supported by experience."

At page 7 of the same volume of the Journal of Health we find the following remarks. The editors were alluding to those persons who think they cannot preserve their health and strength without flesh or fish, and who believe their children would also suffer without it:

"For the information of all such misguided persons, we beg leave to state, that the large majority of mankind do not eat any animal food; or, if any, they use it so sparingly, and at such long intervals, that it cannot be said to form their nourishment. Millions in Asia are sustained by rice alone, with perhaps a little vegetable oil for seasoning.

"In Italy and southern Europe, generally, bread, made of the flour of wheat or Indian corn, with lettuce and the like mixed with oil, constitutes the food of the most robust part of its population.

"The Lazzaroni of Naples, with forms so actively and finely proportioned, cannot even calculate on this much. Coarse bread and potatoes is their chief reliance. Their drink of luxury is a glass of iced water, slightly acidulated.

"Hundreds of thousands—we might say millions—of Irish do not see flesh-meat or fish from one week's end to another. Potatoes and oatmeal are their articles of food: if milk can be added it is thought a luxury. Yet where shall we find a more healthy and robust population, or one more enduring of bodily fatigue, and exhibiting more mental vivacity? What a contrast between these people and the inhabitants of the extreme north—the timid Laplanders, Esquimaux, and Samoideans, whose food is almost entirely animal?"

Again, at page 187 we are told that "the more simple the aliment, and the less *altered* by culinary processes, the slower is the change in digestion; but, at the same time, the less is the stimulation and wear of the powers of life. The Bramins of Hindostan, who live on exceedingly simple food, are long livers, even in a hot and exhausting climate. The peasants of Switzerland and of Scotland, nourished on bread, milk, and cheese, attain a very old age, and enjoy great bodily strength.

"Where there is too much excitement of the body, generally, from fullness of the blood-vessels, or of any one of the organs, owing to a wrong direction of the blood to it (and in one or the other of these conditions we find almost every body now-a-days), animal food, by being long retained in the stomach, and calling into greater action other parts during digestion, as well as furnishing them with more blood afterward, must be obviously improper. The more of this kind of food is taken under such circumstances, the greater will be the oppression; and the weakness, different from that of a healthy person long hungered, will only be increased by the increased amount of blood carried to the diseased part."

It is true that the editors of the Journal of Health connect with the foregoing paragraphs the statement that, "if it be desirable to give nutriment in a small bulk, to obtund completely the sensation of hunger and restore strength to the body, a small quantity of animal will be preferable to much vegetable food." But then it is only in a few diseased cases that any such thing is desirable. And even then, if we look carefully at the language used, the comparison is not made between animal and vegetable food in moderate or reasonable quantities, but between a *small quantity* of the former and *much* of the latter.

DR. J. V. C. SMITH, OF BOSTON.

The following remarks are extracted from the Boston Medical Intelligencer, at a period when Dr. J. V. C. Smith was the editor. They have the appearance of being from Dr. Smith's own pen. Dr. S. is at present the editor of the Boston Medical and Surgical Journal:

"It is true[13] that animal food contains a greater portion of nutriment, in a given quantity, than vegetables; but the digestive functions of the human system become prematurely exhausted by constant action, and the whole system eventually sinks under great or uninterrupted excitement. If, for the various ragouts with which modern tables are so abundantly furnished, men would substitute *wholesome vegetables and pure water*, we should see health walking in paths that are now crowded with the bloated victims of voluptuous appetite. Millions of Gentoos have lived to an advanced age without having tasted any thing that ever possessed life, and been wholly free from a chain of maladies which have scourged every civilized nation on the globe. The wandering Arabs, who have traversed the barren desert of Sahara, subsisting on the scanty pittance of milk from the half-famished camel that carried them, have seen two hundred years roll round without a day of sickness."

SYLVESTER GRAHAM.

Although Mr. Graham does not, so far as I know, lay claim to the "honors" of any medical institution, it cannot be doubted that his knowledge of physiology, to say nothing of anatomy, pathology, and medicine, is such as to entitle him to a high rank among medical men; and I have, therefore, without hesitation, concluded to insert his testimony in this place.

Of his views, however, on the subject before us, it seems almost superfluous to speak, as they are set forth, and have been set forth for many years, so conspicuously, not only in his public lectures, but in his writings, that the bare mention of his name, in almost any part of the country, is to awaken the prejudices, if not the hostilities, of every foe, and of some friends (supposed friends, I mean), of "temperance in all things." It is sufficient, perhaps, for my present purpose, to say of him, that, after the most rigid and profound examination of the subject which he is capable of making—and his capabilities are by no means very limited—it is his unhesitating belief, that in every climate, and in all circumstances in which it is proper for man to be placed, an exclusively farinaceous and fruit diet is the best adapted to the development and improvement of all his powers of body, mind, and soul; provided, however, he were trained to it from the first. And even at any period of life, unless in the case of certain forms of diseases, he believes it would be preferable to exchange, in a proper manner, every form of mixed diet for one purely vegetable. Such opinions as these, as a part of his views in relation to the physical duties of man, he publicly, and strenuously, and eloquently, announces and defends.

DR. JOHN M. ANDREW.

Dr. Andrew is a practitioner of medicine in Remsen, Oneida county, State of New York. His letter was intended for chapter iv., but came too late. This fact is the only apology for inserting it in this place. Several interesting cases of dietetic reform accompanied the letter, but I must omit them, for want of room, in this work.

REMSEN, April 28, 1838.

DEAR SIR—It is now about sixteen months since I adopted an exclusively vegetable diet. I have, however, never been very much inclined to animal food; and, indeed, before I ever heard of the Graham system I laid it aside, during summer, when farming—which, by the by, had always been my occupation till I commenced my professional course, about four years ago. I have, to the best of my knowledge, enjoyed what is commonly called good health, and possessed a degree of strength surpassed only by few; and in connection with the assiduous cultivation of my mental faculties, I have carefully sought to improve my physical powers, which I deem of incalculable worth to the student, as well as to the laborer.

My attention was first called to the subject of vegetable eating by Professor Mussey, in a lecture before the medical class of the Western Medical College of New York, while fulfilling the duties of the professorship, to which he was called in 1836. In that lecture our adaptations, and the design of the Creator in regard to our mode of subsistence, were clearly held forth, and such was the impression made on my mind, that I was induced at once to adopt the vegetable system, both in practice and theory. In my change of diet I did not suffer any inconvenience. The fact that I had, for some length of time, been living mostly on vegetables, will account for that circumstance, however.

But the great advantages derived from the change were soon perceptible, though not appreciated by others. I met with much opposition from my friends, frequently being told that I was fast losing my flesh and all my youthful vigor and vivacity. And yet, for one year and more, I have not lost a pound of flesh.

I was gazed upon as an anomaly in society; some anxiously looking, and others fearfully expecting my downfall and destruction; but both are alike disappointed. The system, though I have not been able to follow it so strictly as I could wish, from the circumstances in which I have been placed, has far exceeded my expectations. One year and more has rolled away, and I thank God I can look back, with some degree of satisfaction, on the time spent in the enjoyment of that alone which sweetens the cup of life. My most able advocacy has been my manual exertions and I have demonstrated the utility of the *system* alike to the professional and laboring classes of community.

I do not go beyond the truth when I say, that I cannot find a man to vie with me in the field, with the scythe, the fork, or the axe. I do not want any thing but potatoes and salt; and I can cut and put up four cords of wood in a day, with no very great exertion. I have frequently been told, by friends, that my *potato and salt system* would not stand the test of the field; but I have silenced their clamor by actual demonstration with all the implements above named.

At present, no consideration would induce me to return to my former mode of living.

JOHN M. ANDREW.

DR. WILLIAM SWEETSER, OF BOSTON.

Dr. Sweetser is the author of a "Treatise on Consumption," and of a "Treatise on Digestion." He has also been a medical professor in the University of Vermont, and a public lecturer on health, in Boston.

In his work on consumption, while speaking of the prevailing belief of a necessity for the use of animal food to those children who possess the scrofulous or consumptive tendency, he thus remarks:

"A diet of milk and mild farinaceous articles, with perhaps light animal decoctions, appears best suited to the early years of life. Whenever there exists an evident inflammatory tendency, as is the case in some scrofulous systems, solid animal food, if used at all, should be taken with the greatest precaution.

"And again—how often is it that fat, plethoric, meat-eating children, their faces looking as though the blood was just ready to ooze out, are with the greatest complacency exhibited by their parents as patterns of health! But let it ever be remembered, that the condition of the system popularly called rude or full health, and which is the result of high feeding, is too often closely bordering on a state of disease."

In his work on digestion he seems to regard man as naturally an omnivorous animal; and, taking this for granted, he speaks as follows respecting his diet:

"One would hardly assert that even in temperate climates his (man's) system requires animal food. I doubt whether any instance can be adduced—unless man be regarded as such—of an omnivorous animal incapable of being adequately nourished by a sufficient and proper vegetable diet.

"Man, dwelling in a temperate climate, and with the power to choose, almost uniformly employs a mixture of animal and vegetable food; but how much early education may have to do in forming his taste for a mixed diet it is difficult to estimate. Habit has certainly great influence in attaching us to particular kinds of aliment. One who has long been accustomed to animal food cannot at once abstain from it without experiencing some feebleness for the want of its stimulation, and perhaps even temporary emaciation. And, on the other hand, he who has long been confined to a vegetable diet is apt to lose his relish for flesh, and, on recurring suddenly to its use, to find it too exciting.

"The liberal use of animal food has been generally thought requisite in arctic climes, to stimulate the functions, and thus furnish a more abundant supply of animal heat, to preserve against the extremity of external temperature. Northern voyagers mostly believe that fat animal food and oils are essential to the maintenance of health and life in the inhabitants of those frozen regions. But to me it would seem that their habits, in respect to diet, prove the *capabilities*, rather than the necessities, of their systems. They learn to eat their coarse fare because they can get no other. Their food, moreover, as is generally the case in savage life, is precarious; and thus, being at times exposed to extreme want, they are stimulated to greater excesses when their supplies are ample.

"The fact of man's dwelling in them (the arctic regions), and eating what he can get there, no more proves him to be naturally a flesh-eating animal than the circumstance of some cattle learning to eat fish, when they are in situations where they can obtain no other food, proves them to be piscivorous.

"Haller conceived it necessary that human life should be sustained by animal and vegetable food, so apportioned that neither should be in excess; and he asserts that abstinence from animal food causes great weakness in the body, and usually a troublesome diarrhœa. But such an opinion is certainly incorrect, since not only particular individuals, but even numbers of people, dwelling in temperate climates, from various causes, subsist almost wholly on vegetable substances, and yet preserve their health and vigor.

"Were we educated to its exclusive use, I am persuaded that a vegetable diet would afford us ample support; but whether, if restrained from animal food, we should, *as a consequence*, in the course of time, and under equally favoring circumstances in other respects, rise still higher in our moral and physical nature, remains, as I conceive, to be proved."

These views of Dr. S. were repeated, in substance, in a course of lectures given by him at the Masonic Temple, in Boston, in 1838. It will be seen that he concedes what the friends of the vegetable system deem a very important point, viz., that man's whole powers, physical, intellectual, and moral, can be well developed on a diet exclusively vegetable. We do not ask him to grant more. If man is as well off on vegetable food as without it, we have moral reasons of so much weight to place against animal food, as, when duly considered, will be, by all candid persons, sufficient to lead to its rejection.

True, we do not believe, with Dr. S.—at least I do not—that "whether a diet purely vegetable, or one comprehending both animal and vegetable food, would be most conducive to health, longevity, and intellectual, moral, and physical development, is a question only to be determined by a long course of experiments, made by various individuals in equal health, and placed, in all other respects, under as nearly similar circumstances as practicable." I believe this course of experiment does not remain *to be* made, but that it has been made, most fully, during the last four or five thousand years, and that the question is settled in favor—wholly so—of vegetable food. Still I do not ask physicians and other medical men to grant more than Dr. S. has; it is quite as much as we ought to expect of them.

DR. A. L. PIERSON.

Dr. Pierson, of Salem, in Massachusetts, a physician and surgeon of considerable eminence, in a lecture some time ago, before the American Institute of Instruction, observed that "young men who were anxious to avail themselves of the advantages of a liberal education, and were therefore compelled to consult economy, had found out that it was not necessary to pay three or four dollars a week for mere board, when the most vigorous and uniform health may be secured by a diet of mere vegetable food and water."

I know not that Dr. P. avows himself an advocate for the exclusive use of vegetable food, but if what I have quoted is not enough to satisfy us in regard to his opinion of its safety, and its full power to develop body and mind, I know not what would be. If the most vigorous and uniform health can be secured on vegetable food, what individual in the world—in view of the moral considerations at least—would ever resort to the carcasses of animals?

STATEMENT OF DR. C. BYINGTON, OF PHILADELPHIA.

A physician of some eminence, residing in Philadelphia, has been heard to say that it was his decided opinion that mankind would live longest, and be healthiest and happiest, on mere bread and water. I may add here, that there was every evidence but one that he was sincere in this statement, although I do not fully accord with him, believing that the best health requires variety of food—not, indeed, at the same meal, but at different ones. The exception I make in regard to his sincerity, is in reference to the fact, that while he professed to believe a bread and vegetable diet to be best for mankind, he did not adopt it.

TESTIMONY OF A PHYSICIAN IN NEW YORK.

In the work entitled "Hints to a Fashionable Lady," by a physician—his name not given—we find the following testimony:

"Young persons invariably do best on simple but moderately nutritious fare. Too large a proportion of animal food and fatty substances are pernicious to the complexion. On the contrary, a diet which is principally vegetable, with the luxuries of the dairy (not butter, surely, for that is elsewhere prohibited), is most advantageous. Nowhere are finer complexions to be found than in those parts of England, Scotland, and Ireland, where the living is almost exclusively vegetable.

"Those who subsist entirely on vegetable food have seldom, if ever, a constantly bad breath, or an offensive perspiration. It has been ascertained that the teeth are uniformly best in those countries where least animal food is used."

THE FEMALE'S CYCLOPEDIA.

From a fugitive volume, entitled "The Female's Cyclopedia," I have concluded to make the following extract, because I have reason to believe the writer to have been a physician:

"Animal food certainly gives most strength; but its stimulancy excites fever, and produces plethora and its consequences. The system is sooner worn out by a repetition of its stimuli, and those who indulge greatly in such diet are more likely to be carried off early by inflammatory diseases; or if, by judicious exercise, they qualify its effects, they yet acquire such an accumulation of putrescent fluids as becomes the foundation for the most inveterate chronic diseases in after age.

"The most valuable state of the mind, however, appears to be connected with somewhat less of firmness and vigor of body. Vegetable aliment, as never over-distending the vessels or loading the system, does not interrupt the stronger emotions of the mind; while the heat, fullness, and weight of animal food, are inimical to its vigorous exertion. Temperance, therefore, does not so much consist in the quantity—since the appetite will regulate that—as in the quality; namely, in a large proportion of vegetable aliment."

DR. VAN COOTH.

Dr. Van Cooth, a learned European writer—I believe a Hollander—has recently maintained, incidentally, in a learned medical dissertation, that the great body of the ancient Egyptians and Persians "confined themselves to a vegetable diet." To be sure, Dr. V. does not seem to be a vegetable eater himself, but the friends of the latter system are not the less indebted to him for the concession. The physical and moral superiority of those vegetable eating nations, in the days of their glory, are well known; and every intelligent reader of history, and honest inquirer after truth, will make his own inferences from the facts which I have mentioned.

DR. WILLIAM BEAUMONT.

The work of this gentleman, entitled "Experiments and Observations on the Gastric Juice, and the Physiology of Digestion," is well known—at least to the medical community. The following are some of the conclusions to which his experiments conducted him:

"Solid aliment, thoroughly masticated, is far more salutary than soups, broths, etc.

"Fat meats, butter, and oily substances of every kind, are difficult of digestion, offensive to the stomach, and tend to derange that organ and induce disease.

"Spices, pepper, stimulating and heating condiments of every kind, retard digestion and injure the stomach.

"Coffee and tea debilitate the stomach and impair digestion.

"Simple water is the only fluid called for by the wants of the economy; the artificial drinks are all more or less injurious—some more so than others; but none can claim exemption from the general charge."

If it should be said that this testimony of Dr. Beaumont is by no means directly in favor of a diet exclusively vegetable. I admit it. But he certainly goes very far toward conceding every thing which I claim, when he says that "fat meats, butter, and oily substances of every kind, are difficult of digestion, offensive to the stomach, and tend to derange that organ and induce disease;" and especially when he speaks so highly of farinaceous substances and good fruits. Pray, what animal food can be eaten which does not contain, at least, a small quantity of oil? And if this oil tends to induce disease, and farinaceous food does not, why should not animal food be excluded?

SIR EVERARD HOME.

This distinguished philosopher and medical gentleman, though, like many others, he insisted that vegetable food did not produce full muscular development, yet admitted the natural character of man to be that of a vegetable eater, in the following, or nearly the following, terms:

"In the history of man—in the Bible—we are told that dominion over the animal world was bestowed upon him at his creation; but the divine permission to indulge in animal food was not given till after the flood. The observations I have to make accord strongly with this tradition; for, while mankind remained in a state of innocence, there is every ground to believe that their only food was the produce of the vegetable kingdom."

DR. JENNINGS.

Dr. Jennings is the author of a work published at Oberlin, Ohio, in 1847, entitled "Medical Reform." In this volume, at page 198, we find the following facts and statements. The author is comparing the effects of animal food on the human system with those of alcohol, from which we learn his views concerning the former:

"Position I.—Animal food, in common with alcohol, creates a feverish diathesis, evidences of which are—1. An impaired state of the respiratory function. 2. The pulse is rendered more frequent and irregular, both by alcohol and meat. 3. A feverish heat is generated in the system, and persons are made more thirsty, by the use of both these substances. 4. Both substances equally induce what is called the digestive fever.

"Position II.—Alcoholic drinks lay the foundation for occasional disturbances in the system, of different kinds and grades, as bilious bowel affections, etc., and so do flesh meats. In the production of colds, animal food is far the most efficient.

"Position III.—Animal food tends, quite as strongly as the moderate use of alcoholic liquors, to weaken and disturb the balance of action between the secerning and excerning systems of vessels, by which some persons become leaner and others fleshier than they should be.

"Position IV.—With about equal potency alcohol and flesh meats weaken the force of the capillaries of the system, on which healthy action so much depends.

"Position V.—A flesh diet, in common with the use of strong drink, impairs the tone of the nutritive apparatus, by which its ability to work up raw material and manufacture it into sound, well finished vital fabric, is diminished, and of course the appetite or call for food is satisfied with a less quantity of the raw material. This fact has given rise to the opinion that animal food contains more nutriment than vegetable.

"Position VI.—The total abandonment of an habitual use of animal food is attended with all the perplexing, uncomfortable, and distressing difficulties that follow the giving up of an habitual use of strong drink. A change from one kind of simple nutriment to another has no such effect. It is only when the constant use of some stimulating substance is abandoned that such difficulties are experienced."

DR. JARVIS.

This gentleman, in his "Practical Physiology," at page 86, has the following thoughts:

"Some have contended that man was designed to eat only of the fruits and vegetables of the earth; while others maintain, with equal confidence, that he should add to these the flesh of beasts. There are many individuals, both in this and other countries, who confine themselves to vegetable diet. They believe they enjoy better health, and maintain greater strength of body and mind, than those who live on a mixed diet. The experiment has not been tried on a sufficiently extensive range to determine its value. It has not proved a failure, nor has it demonstrated, to the satisfaction of all, that flesh is injurious."[14]

DR. TICKNOR.

"From the fact," says this author, "that animal food is proper and necessary for health in polar regions, and that a vegetable diet is equally proper and necessary in the torrid zone, we may conclude that in winter, in our own climate, an animal diet is the best; while vegetables are more conducive to health in the summer season."

It would not be difficult to prove, from the very concessions of Dr. T., that vegetable food is better adapted to health, in *general*, than animal; but I forbear to do so, in this place. The subject will be fully discussed in the concluding chapter.

DR. COLES.

The author of a small volume recently published at Boston, entitled the "Philosophy of Health; or, Health without Medicine," is more decided in his views on diet than any late writer I have seen, except Dr. Jennings and O. S. Fowler. He says, at page 35:

"Man, in his original, holy state, was provided for from the vegetables of that happy garden which was given him to prune. This was the Creator's original plan; * * * * the eating of flesh was one of the consequences of the fall. Living on vegetable food is undoubtedly the most natural and healthy method of subsistence."

Again, at page 45—"The objections, then, against meat-eating are threefold—intellectual, moral, and physical. Its tendency is to check intellectual activity, to depreciate moral sentiment, and to derange the fluids of the body."

DR. SHEW.

This active physician is zealously devoted to the propagation of hydropathy. He uses no medicine in the management of disease—nothing at all but water. To this, however, he adds great attention to diet. In his Journal,[15] and elsewhere, he is a zealous and able advocate of the vegetable system, preferring it himself, and recommending it to his patients and followers.

Dr. Shew's opinion, in this particular, is entitled to the more weight from the fact of his having been very familiar with disease and diet, both in the old world and the new. He has been twice to Germany; and has spent much time at Graefenberg, with Priessnitz, the founder of the system which he so zealously defends and practices, and so strongly advocates.

DR. MORRILL.

Dr. C. Morrill, in a recent work entitled, "Physiology of Woman, and her Diseases," says much in favor of an exclusively vegetable diet in some of the diseases of woman; and among other things, makes the following general remarks:

"Even by those who labor (referring here to the healthy), meat should be taken moderately, and but once a day. The sedentary, generally, do not need it."

DR. BELL.

This gentleman's testimony has been given elsewhere. I only subjoin the following: "By far the greater number of the inhabitants of the earth have used, in all ages, and continue to use, at this time, vegetable aliment alone."

DR. BRADLEY.

Dr. D. B. Bradley, the distinguished missionary at Bangkok, in Siam, though not exactly a vegetable eater, is favorably disposed to the vegetable system. He has read Graham and myself with great care, and is an anxious inquirer after all truth.

DR. STEPHENSON.

Dr. Chauncy Stephenson, of Chesterfield, Massachusetts, in what he calls his "New System of Medicine," commends to all his readers, for their sustenance, "pure air, a proper temperature, good vegetable food, and pure cold water." And lest he should be misunderstood, he immediately adds— "The best articles of food for general use are good, well-baked cold bread, made of rye and Indian corn, wheat or barley meal; rice, good ripe fruits of all kinds, both fresh and dried, and a proper proportion of good roots, such as potatoes, parsneps, turnips, onions, etc." Even milk he regards as a questionable food for adults or middle aged persons.

Again, he says: "Animal food, in general, digests sooner than most kinds of vegetables; and not being so much in accordance with man's nature, constitution, and moral character, it is very liable, finally, to generate disease, inflammation, or fever, even when it is not taken to excess." He closes by advising all persons to content themselves with "pure vegetable food;" and that in the least quantity compatible with good health.

DR. J. BURDELL,

A distinguished dentist of New York, has long been a vegetable eater, and a zealous defender of the faith (in this particular) which he professes.

DR. THOMAS SMETHURST,

In a work entitled Hydrotherapia, says, "Children thrive best upon a simple, moderately nourishing vegetable diet." And if children thus thrive the best, why not adults?

DR. SCHLEMMER.

Dr. C. V. Schlemmer, a German by birth, but now an adopted son of old England, in giving an account of the diet of himself, his three sons of eleven, ten, and four years of age, with their tutor, observes: "Raw peas, beans, and fruit are our food: our teeth are our mills; the stomach is the kitchen." And all of them, as he affirms, enjoy the best of health. For himself, as he says, he has practiced in this way six years.

DR. CURTIS, AND OTHERS.

Dr. Curtis, a distinguished botanic physician of Ohio, with several other physicians, both of the old and the new school, whom I have not named, do not hesitate to regard a pure vegetable diet, in the abstract, as by far the best for all mankind, both in health and disease.

Dr. Porter, of Waltham, for example, when I meet him, always concedes that a well-selected vegetable diet is superior to every other. He has repeatedly told me of an experiment he made, of three months, on mere bread and water. Never, says he, was I more vigorous in body and mind, than at the end of this experiment. But the reader well knows that I am not an advocate of a diet of mere bread and water. I regard fruits, or fruit juices—unfermented—almost as necessary, to adults, as bread.

PROF. C. U. SHEPARD.

The reputation of this gentleman, in the scientific world, is so well known, that no apology can be necessary for inserting his testimony. As a chemist, he is second to very few, if any, men in this country. The following are his remarks:

"Start not back at the idea of subsisting upon the potato alone, ye who think it necessary to load your tables with all the dainty viands of the market—with fish, flesh, and fowl, seasoned with oil and spices, and eaten, perhaps, with wines;—start not back, I say, with disgust, until you are able to display in your own pampered persons a firmer muscle, a more beau-ideal outline, and a healthier red than the potato-fed peasantry of Ireland and Scotland once showed you, as you passed by their cabin doors!

"No; the chemical physiologist will tell you that the well ripened potato, when properly cooked, contains every element that man requires for nutrition; and in the best proportion in which they are found in any plant whatever. There is the abounding supply of starch for enabling him to maintain the process of breathing, and for generating the necessary warmth of body; there is the nitrogen for contributing to the growth and renovation of organs; the lime and phosphorus for the bones; and all the salts which a healthy circulation demands. In fine, the potato may well be called the universal plant."

BLACKWOOD, IN HIS MAGAZINE.

"Chemistry," says Blackwood's Magazine, "has already told us many remarkable things in regard to the vegetable food we eat—that it contains, for example, a certain per centage of the actual fat and lean we consume in our beef, or mutton, or pork—and, therefore, that he who lives on vegetable food may be as strong as the man who lives on animal food, because both in reality feed on the same things, in a somewhat different form."

There is this difference, however, that in the one case—that is, in the use of the vegetables which contain the elements referred to—we save the trouble of running it through the body of the living animal, and losing seven eighths of it, as we do, practically in the process; whereas in the other we do not. We also save ourselves the necessity of training the young and the old to scenes of butchery and blood.

PROF. JOHNSTON.

This gentleman, in a recent edition of his "Elements of Agricultural Chemistry and Geology," tells us that from experiments made in the laboratory of the Agricultural Association of Scotland, wheat and oats, when analyzed, contain of nutritious properties the following proportion:

	Musc. matter.	Fat.	Starch.
Wheat,	10 pounds,	3 pounds,	50 pounds.
Oats,	18 "	6 "	65 "

Thus oats, and even wheat, are quite rich in that which forms muscular matter in the human body.

SIMEON COLLINS, OF WESTFIELD, MASS.

This gentleman, in his fifty-first year, states that having been for several years afflicted with a severe cough, which he supposed bordered upon consumption, he "discontinued the use of flesh meat, fish, fowl, butter, gravy, tea, and coffee, and made use of a plain vegetable diet." "My bread," says he, "is made of unbolted wheat meal; my drink is pure cold water; my bed, for winter and summer, is made of the everlasting flower; and my health is, and ever has been, perfect, since I got fairly cleansed from the filthiness of flesh meat, and other pernicious articles of diet in common use.

"My business requires a great degree of activity, and I can truly say that I am a stranger to weariness or languor. At the time of entering upon this system, I had a wife and five children, the youngest eight years of age;—they all soon entered upon the same course of living with myself, and soon were all benefited in health. I have now six children—the youngest fifteen months old, and as happy as a lark. Previous to the time of our adopting the present system of living, my expenses for medicine and physicians would range from $20 to $30 a year—for the last four years it has been nothing worth naming."

REV. JOSEPH EMERSON.

Mr. Emerson was a teacher of eminence, known throughout the United States, but particularly so in Massachusetts and Connecticut. He died in the latter state, in 1833, aged about fifty-five. He had long been a miserable dyspeptic, but was probably kept alive amid certain strange violations of physical law, such as studying hard till midnight, for example, for many years, by his great care in regard to his diet. Mrs. Banister, late Miss Z. P. Grant (the associate, at Ipswich, of Miss Lyon, who died recently at South Hadley, who was his pupil), thus speaks of his rigid habits:

"He not only uniformly rejected whatever food he had decided to be injurious to him, but whatever he deemed necessary for his food or drink, was always taken, whether at home or abroad. As his diet, for several years, consisted generally, either of bread and milk, or of bread and butter, what solid food he wanted could be supplied at any table."[16]

It is also testified of him, by his brother, Prof. Emerson, of Andover, that "for more than thirty years he adopted the practice of eating but one kind at a meal." If I do not misremember, for I knew him well, he was in favor of banishing flesh and fish, and substituting milk and fruits in their stead, on Bible ground.—I refer here to the Divine arrangement in the first chapter of Genesis; and which has never, that I am aware, been altered.

TAK SISSON.

Tak Sisson, as he was called, was a slave in the family of a man in Rhode Island, before and during the Revolution.

From early childhood he could never be prevailed on to eat any flesh or fish, but he subsisted on vegetable food and milk; neither could he be persuaded to eat high seasoned food of any kind. When he was a child, his parents used to scold him severely, and threaten to whip him because he refused to eat flesh. They said to him (as I have been told a thousand times), that if he did not eat meat he would never be good for any thing, but would always be a poor, puny creature.

But Tak persevered in his vegetable and unstimulating diet, and, to the surprise of all, grew fast, and his body was finely developed and athletic. He was very stout and robust, and altogether the most vigorous and dexterous of any of the family. He finally became more than six feet high, and every way well proportioned, and remarkable for his agility and strength. He was so uncommonly shrewd, bright, strong, and active, that he became notorious for his shrewdness, and for his feats of strength and agility. Indeed, he was so full of his playful mischief as greatly to annoy his overseer.

During the Revolutionary War it became an object to take Gen. Prescott. A door was to be forced where he was quartered and sleeping, and Tak was selected for the work. Having taken his lesson from the American officer, he proceeded to the door, plunged his thick head against it, burst it open, roused Gen. P., like a tiger sprung upon him, seized him in his brawny arms, and in a low, stern voice, said, "One word, and you are a dead man." Then hastily snatching the general's cloak and wrapping it round him, at the same time telling a companion to take care of the rest of his clothes, he took him in his arms, as if a child, and ran with him to a boat which was waiting, and escaped with his prisoner without rousing even the British sentinels.

Tak lived on his vegetable fare to a very advanced age, and was remarkable, through life, for his activity, strength, and shrewdness.

FOOTNOTES:

[9] By seed, Dr. C. means the farinaceous grains; wheat, corn, rye, etc.

[10] Cuvier was not a medical man, but I have classed him with medical men, on account of his profound knowledge of Comparative Anatomy and Physiology.

[11] "Unless," as a writer in the Graham Journal very justly observes, "these latter indulge, habitually and freely, in the use of intoxicating substances."

[12] Such was Gen. Elliot, so distinguished at the famous siege of Gibraltar. Such, too, was Mr. Shillitoe, of whom honorable mention will be made in another place;—besides many more.

[13] So he thinks, but I think otherwise. Animal food, as I have shown elsewhere, is not so nutritious as some of the farinaceous vegetables.

[14] Dr. J. here overlooks one important fact, viz., that the testimony of all those who have tried the exclusive use of vegetable food is *positive* in its nature; while that of others, who have not tried it, is, and necessarily must be, negative.

[15] The Water-Cure Journal.

[16] An aged lady, of Dedham—a pillar in every good cause—has, for twelve or fifteen years, carried abroad with her, when traveling, some plain bread and apples; and no entreaties will prevail with her, at home or abroad, to eat luxuries.

CHAPTER VI.

TESTIMONY OF PHILOSOPHERS AND OTHER EMINENT MEN.

General Remarks.—Testimony of Plautus.—Plutarch.—Porphyry.—Lord Bacon.—Sir William Temple.—Cicero.—Cyrus the Great.—Gassendi.—Prof. Hitchcock.—Lord Kaims.—Dr. Thomas Dick.—Prof. Bush.—Thomas Shillitoe.—Alexander Pope.—Sir Richard Phillips.—Sir Isaac Newton.—The Abbé Gallani.—Homer.—Dr. Franklin.—Mr. Newton.—O. S. Fowler.—Rev. Mr. Johnston.—John H. Chandler.—Rev. J. Caswell.—Mr. Chinn.—Father Sewall.—Magliabecchi.—Oberlin and Swartz.—James Haughton.—John Bailies.—Francis Hupazoli.—Prof. Ferguson.—Howard, the Philanthropist.—Gen. Elliot.—Encyclopedia Americana.—Thomas Bell, of London.—Linnæus, the Naturalist.—Shelley, the Poet.—Rev. Mr. Rich.—Rev. John Wesley.—Lamartine.

GENERAL REMARKS.

This chapter might have been much more extended than it is. I might have mentioned, for example, the cases of Daniel and his three brethren, at the court of the Babylonian monarch, who certainly maintained their health—if they did not even improve it—by vegetable food, and by a form of it, too, which has by many been considered rather doubtful. I might have mentioned the case of Paul,[17] who, though he occasionally appears to have eaten flesh, said, expressly, that he would abstain from it while the world stood, where a great moral end was to be gained; and no one can suppose he would have done so, had he feared any injury would thereby result to his constitution of body or mind.

The case of William Penn, if I remember rightly what he says in his "No Cross no Crown," would have been in point. Jefferson, the third President of the United States, was, according to his own story, almost a vegetable eater, during the whole of his long life. He says he abstained principally from animal food; using it, if he used it at all, only as a condiment for his vegetables. And does any one, who has read his remarks, doubt that his "convictions" were in favor of the exclusive use of vegetable food?

However, to prevent the volume from much exceeding the limits originally assigned it, I will be satisfied—and I hope the public will—with the following selections of testimonies, ancient and modern; some of more, some of less importance; but all of them, as it appears to me, worthy of being collected and incorporated into a volume like this, and faithfully and carefully examined.

PLAUTUS.

Plautus, a distinguished dramatic Roman writer, who flourished about two thousand years ago, gives the following remarkable testimony against the use of animal food, and of course in favor of the salubrity of vegetables; addressed, indeed, to his own countrymen and times, but scarcely less applicable to our own:

"You apply the term wild to lions, panthers, and serpents; yet, in your own savage slaughters, you surpass them in ferocity; for the blood shed by them is a matter of necessity, and requisite for their subsistence.

"But, that man is not, by nature, destined to devour animal food, is evident from the construction of the human frame, which bears no resemblance to wild beasts or birds of prey. Man is not provided with claws or talons, with sharpness of fang or tusk, so well adapted to tear and lacerate; nor is his stomach so well braced and muscular, nor his animal spirits so warm, as to enable him to digest this solid mass of animal flesh. On the contrary, nature has made his teeth smooth, his mouth narrow, and his tongue soft; and has contrived, by the slowness of his digestion, to divert him from devouring a species of food so ill adapted to his frame and constitution. But, if you still maintain that such is your natural mode of subsistence, then follow nature in your mode of killing your prey, and employ neither knife, hammer, nor hatchet—but, like wolves, bears, and lions, seize an ox with your teeth, grasp a boar round the body, or tear asunder a lamb or a hare, and, like the savage tribe, devour them still panting in the agonies of death.

"We carry our luxury still farther, by the variety of sauces and seasonings which we add to our beastly banquets—mixing together oil, wine, honey, pickles, vinegar, and Syrian and Arabian ointments and perfumes, as if we intended to bury and embalm the carcasses on which we feed. The difficulty of digesting such a mass of matter, reduced in our stomachs to a state of liquefaction and putrefaction, is the source of endless disorders in the human frame.

"First of all, the wild, mischievous animals were selected for food; and then the birds and fishes were dragged to slaughter; next, the human appetite directed itself against the laborious ox, the useful and fleece-bearing sheep, and the cock, the guardian of the house. At last, by this preparatory discipline, man became matured for human massacres, slaughters, and wars."

PLUTARCH.

"It is best to accustom ourselves to eat no flesh at all, for the earth affords plenty enough of things not only fit for nourishment, but for enjoyment and delight; some of which may be eaten without much preparation, and others may be made pleasant by adding divers other things to them.

"You ask me," continues Plutarch, "'for what reason Pythagoras abstained from eating the flesh of brutes?' For my part, I am astonished to think, on the contrary, what appetite first induced man to taste of a dead carcass; or what motive could suggest the notion of nourishing himself with the flesh of animals which he saw, the moment before, bleating, bellowing, walking, and looking around them. How could he bear to see an impotent and defenceless creature slaughtered, skinned, and cut up for food? How could he endure the sight of the convulsed limbs and muscles? How bear the smell arising from the dissection? Whence happened it that he was not disgusted and struck with horror when he came to handle the bleeding flesh, and clear away the clotted blood and humors from the wounds?

"We should therefore rather wonder at the conduct of those who first indulged themselves in this horrible repast, than at such as have humanely abstained from it."

PORPHYRY, OF TYRE.

Porphyry, of Tyre, lived about the middle of the third century, and wrote a book on abstinence from animal food. This book was addressed to an individual who had once followed the vegetable system, but had afterward relinquished it. The following is an extract from it:

"You owned, when you lived among us, that a vegetable diet was preferable to animal food, both for preserving the health and for facilitating the study of philosophy; and now, since you have eat flesh, your own experience must convince you that what you then confessed was true. It was not from those who lived on vegetables that robbers or murderers, sycophants or tyrants, have proceeded; but from *flesh-eaters*. The necessaries of life are few and easily acquired, without violating justice, liberty, health, or peace of mind; whereas luxury obliges those vulgar souls who take delight in it to covet riches, to give up their liberty, to sell justice, to misspend their time, to ruin their health and to renounce the joy of an upright conscience."

He takes pains to persuade men of the truth of the two following propositions:

1st. "That a conquest over the appetites and passions will greatly contribute to preserve health and to remove distempers.

2d. "That a simple vegetable food, being easily procured and easily digested, is a mighty help toward obtaining this conquest over ourselves."

To prove the first proposition, he appeals to experience, and proves that many of his acquaintance who had disengaged themselves from the care of amassing riches, and turning their thoughts to spiritual subjects, had got rid entirely of their bodily distempers.

In confirmation of the second proposition, he argues in the following manner: "Give me a man who considers, seriously, what he is, whence he came, and whither he must go, and from these considerations resolves not to be led astray nor governed by his passions; and let such a man tell me whether a rich animal diet is more easily procured or incites less to irregular passions and appetites than a light vegetable diet! But if neither he, nor a physician, nor indeed any reasonable man whatsoever, dares to affirm this, why do we oppress ourselves with animal food, and why do we not, together with luxury and flesh meat, throw off the incumbrances and snares which attend them?"

LORD BACON.

Lord Bacon, in his treatise on Life and Death, says, "It seems to be approved by experience, that a spare and almost a Pythagorean diet, such as is prescribed by the strictest monastic life, or practiced by hermits, is most favorable to long life."

SIR WILLIAM TEMPLE.

"The patriarchs' abodes were not in cities, but in open countries and fields. Their lives were pastoral, and employed in some sorts of agriculture. They were of the same race, to which their marriages were generally confined. Their diet was simple, as that of the ancients is generally represented. Among them flesh and wine were seldom used, except at sacrifices at solemn feasts.

"The Brachmans, among the old Indians, were all of the same races, lived in fields and in woods, after the course of their studies was ended, and fed only upon rice, milk, and herbs.

"The Brazilians, when first discovered, lived the most natural, original lives of mankind, so frequently described in ancient countries, before laws, or property, or arts made entrance among them; and so their customs may be concluded to have been yet more simple than either of the other two. They lived without business or labor, further than for their necessary food, by gathering fruits, herbs, and plants. They knew no other drink but water; were not tempted to eat or drink beyond common appetite and thirst; were not troubled with either public or domestic cares, and knew no pleasures but the most simple and natural.

"From all these examples and customs, it may probably be concluded that the common ingredients of health and long life are, great temperance, open air, easy labor, little care, simplicity of diet—rather fruits and plants than flesh, which easier corrupts—and water, which preserves the radical moisture without too much increasing the radical heat. Whereas sickness, decay, and death proceed commonly from the one preying too fast upon the other, and at length wholly extinguishing it."

CICERO.

This eminent man sometimes, if not usually, confined himself to vegetable food. Of this we have evidence, in his complaints about the refinements of cookery—that they were continually tempting him to excess, etc. He says, that after having withstood all the temptations that the noblest lampreys and oysters could throw in his way, he was at last overpowered by paltry beets and mallows. A victory, by the way, which, in the case of the eater of plain food, is very often achieved.

CYRUS THE GREAT.

This distinguished warrior was brought up, like the inferior Persians, on bread, cresses, and water; and, notwithstanding the temptations of a luxurious and voluptuous court, he rigorously adhered to his simple diet. Nay, he even carried his simple habits nearly through life with him; and it was not till he had completely established one of the largest and most powerful empires of antiquity that he began to yield to the luxuries of the times. Had he pursued his steady course of temperance through life, the historian, instead of recording his death at only seventy, might have told us that he died at a hundred or a hundred and fifty.

PETER GASSENDI.

Two hundred and twenty years ago, Peter Gassendi, a famous French philosopher—and by the way, one of the most learned men of his time—wrote a long epistle to Van Helmont, a Dutch chemist, on the question whether the teeth of mankind indicate that they are naturally flesh-eaters.

In this epistle, too long for insertion here,[18] Gassendi maintains, with great ingenuity, that the human teeth were not made for flesh. He does not evade any of the facts in the case, but meets them all fairly and discusses them freely. And after having gone through with all parts of the argument, and answered every other conceivable objection, he thus concludes:

"And here I feel that it may be objected to me: Why, then, do you not, yourself, abstain from flesh and feed only on fruits and vegetables? I must plead the force of habit, for my excuse. In persons of mature age nature appears to be so wholly changed, that this artificial habit cannot be renounced without some detriment. But I confess that if I were wise, and relinquishing the use of flesh, should gradually accustom myself to the gifts of the kind earth, I have little doubt that I should enjoy more regular health, and acquire greater activity of mind. For truly our numerous diseases, and the dullness of our faculties, seem principally produced in this way, that flesh, or heavy, and, as I may say, too substantial food, overloads the stomach, is oppressive to the whole body, and generates a substance too dense, and spirits too obtuse. In a word, it is a yarn too coarse to be interwoven with the threads of man's nature."

I know how it strikes many when they find such men as Gassendi, admitting the doctrines for which I contend, in theory, and even strenuously defending them, and yet setting them at naught in practice. Surely, say they, such persons cannot be sincere. For myself, however, I draw a very different conclusion. Their conduct is perfectly in harmony with that of the theoretic friends of cold water, plain dress, and abstemiousness in general. They are compelled to admit the truth; but it is so much against their habits, as in the case of Gassendi, besides being still more strongly opposed to their lusts and appetites, that they cannot, or rather, will not conform to what they believe, in their daily practice. Their testimony, to me, is the strongest that can be obtained, because they testify against themselves, and in spite of themselves.

PROF. HITCHCOCK.

This gentleman, a distinguished professor in Amherst College, is the author of a work, entitled "Dyspepsia Forestalled and Resisted," which has been read by many, and execrated by not a few of those who are so wedded to their lusts as to be unwilling to be told of their errors.

I am not aware that Professor H. has any where, in his writings, urged a diet exclusively vegetable, for all classes of the community, although I believe he does not hesitate to urge it on all students; and one might almost infer, from his works of various kinds, that if he is not already a believer in the doctrines of its universal superiority to a mixed diet, he is not very far from it. In a sermon of his, in the National Preacher, for November, 1834, he calls a diet exclusively vegetable, a "proper course of living."

I propose to add here a few anecdotes of his, which I know not how to find elsewhere.

"Pythagoras restricted himself to vegetable food altogether, his dinner being bread, honey, and water; and he lived upward of eighty years. Matthew (St. Matthew, I suppose he means), according to Clement, lived upon vegetable diet. Galen, one of the most distinguished of the ancient physicians, lived one hundred and forty years, and composed between seven and eight hundred essays on medical and philosophical subjects; and he was always, after the age of twenty-eight, extremely sparing in the quantity of his food. The Cardinal de Salis, Archbishop of Seville, who lived one hundred and ten years, was invariably sparing in his diet. One Lawrence, an Englishman, by temperance and labor lived one hundred and forty years; and one Kentigern, who never tasted spirits or wine, and slept on the ground and labored hard, died at the age of one hundred and eighty-five. Henry Jenkins, of Yorkshire, who died at the age of one hundred and sixty-nine, was a poor fisherman, as long as he could follow this pursuit; and ultimately he became a beggar, living on the coarsest and most sparing diet. Old Parr, who died at the age of one hundred and fifty-three, was a farmer, of extremely abstemious habits, his diet being solely milk, cheese, coarse bread, small beer, and whey. At the age of one hundred and twenty he married a second wife by whom he had a child. But being taken to court, as a great curiosity, in his one hundred and fifty-second year, he very soon died—as the physicians decidedly testified, after dissection, in consequence of a change from a parsimonious to a plentiful diet. Henry Francisco, of this country, who lived to about one hundred and forty, was, except for a certain period, remarkably abstemious, eating but little, and particularly abstaining almost entirely from animal food; his favorite articles being tea, bread and butter, and baked apples. Mr. Ephraim Pratt, of Shutesbury, Mass., who died at the age of one hundred and seventeen years, lived very much upon milk, and that in small quantity; and his son, Michael Pratt, attained to the age of one hundred and three, by similar means."

Speaking, in another place, of a milk diet, Professor H. observes, that "a diet chiefly of milk produces a most happy serenity, vigor, and cheerfulness of mind—very different from the gloomy, crabbed, and irritable temper, and foggy intellect, of the man who devours flesh, fish, and fowl, with ravenous appetite, and adds puddings, pies, and cakes to the load."

LORD KAIMS.

Henry Home, otherwise called Lord Kaims, the author of the "Elements of Criticism," and of "Six Sketches on the History of Man," has, in the latter work, written eighty years ago, the following statements respecting the inhabitants of the torrid zone:

"We have no evidence that either the hunter or shepherd state were ever known there. The inhabitants at present subsist upon vegetable food, and probably did so from the beginning."

In speaking of particular nations or tribes of this zone, he tells us that "the inhabitants of Biledulgerid and the desert of Sahara, have but two meals a day—one in the morning and one in the evening;" and "being temperate," he adds, "and strangers to the diseases of luxury and idleness, they generally live to a great age."[19] Sixty, with them, is the prime of life, as thirty is in Europe. "Some of the inland tribes of Africa," he says, "make but one meal a day, which is in the evening." And yet "their diet is plain, consisting mostly of rice, fruits, and roots. An inhabitant of Madagascar will travel two or three days without any other food than a sugar-cane." So also, he might have added, will the Arab travel many days, and at almost incredible speed, with nothing but a little gum-arabic; and the Peruvians and other inhabitants of

South America, with a little parched corn. But I have one more extract from Lord Kaims:

"The island of Otaheite is healthy, the people tall and well made; and by temperance—vegetables and fish being their chief nourishment—they live to a good old age, with scarcely an ailment. There is no such thing known among them as rotten teeth; the very smell of wine or spirits is disagreeable; and they never deal in tobacco or spiceries. In many places Indian corn is the chief nourishment, which every man plants for himself."

DR. THOMAS DICK.

Dr. Dick, author of the "Philosophy of Religion," and several other works deservedly popular, gives this remarkable testimony:

"To take the life of any sensitive being, and to feed on its flesh, appears incompatible with a state of innocence, and therefore no such grant was given to Adam in paradise, nor to the antediluvians. It appears to have been a grant suited only to the degraded state of man, after the deluge; and it is probable that, as he advances in the scale of moral perfection in the future ages of the world, the use of animal food will be gradually laid aside, and he will return again to the productions of the vegetable kingdom, as the original food of man—as that which is best suited to the rank of rational and moral intelligence. And perhaps it may have an influence, in combination with other favorable circumstances, in promoting health and longevity."

PROFESSOR GEORGE BUSH.

Professor Bush, a writer of some eminence, in his "Notes on Genesis," while speaking of the permission to man in regard to food, in Genesis i. 29, has the following language:

"It is not perhaps to be understood, from the use of the word *give*, that a *permission* was now granted to man of using that for food which it would have been unlawful for him to use without that permission; for, by the very constitution of his being, he was made to be sustained by that food which was most congenial to his animal economy; and this it must have been lawful for him to employ, unless self-destruction had been his duty. The true import of the phrase, therefore, doubtless is, that God had *appointed, constituted, ordained* this, as the staple article of man's diet. He had formed him with a nature to which a vegetable aliment was better suited than any other. It cannot perhaps be inferred from this language that the use of flesh-meat was absolutely forbidden; but it clearly implies that the fruits of the field were the diet most adapted to the constitution which the Creator had given."

THOMAS SHILLITOE.

Mr. Shillitoe was a distinguished member of the Society of Friends, at Tottenham, near London. The first twenty-five years of his life were spent in feeble health, made worse by high living. This high living was continued about twenty years longer, when, finding himself fast failing, he yielded to the advice of a medical friend, and abandoned all drinks but water, and all food but the plainest kinds, by which means he so restored his constitution that he lived to be nearly ninety years of age; and at eighty could walk with ease from Tottenham to London, six miles, and back again. The following is a brief account of this distinguished man, when at the age of eighty, and nearly in his own words:

It is now nearly thirty years since I ate fish, flesh, or fowl, or took fermented liquor of any kind whatsoever. I find, from continued experience, that abstinence is the best medicine. I don't meddle with fermented liquors of any kind, even as medicine. I find I am capable of doing better without them than when I was in the daily use of them.

"One way in which I was favored to experience help in my willingness to abandon all these things, arose from the effect my abstinence had on my natural temper. My natural disposition is very irritable. I am persuaded that ardent spirits and high living have more or less effect in tending to raise into action those evil propensities which, if given way to, war against the soul, and render us displeasing to Almighty God."

ALEXANDER POPE.

Pope, the poet, ascribes all the bad passions and diseases of the human race to their subsisting on the flesh, blood, and miseries of animals. "Nothing," he says, "can be more shocking and horrid than one of our kitchens, sprinkled with blood, and abounding with the cries of creatures expiring, or with the limbs of dead animals scattered or hung up here and there. It gives one an image of a giant's den in romance, bestrewed with the scattered heads and mangled limbs of those who were slain by his cruelty."

SIR RICHARD PHILLIPS.

Sir Richard Phillips, in his "Million of Facts," says that "the mixed and fanciful diet of man is considered as the cause of numerous diseases, from which animals are exempt. Many diseases have abated with changes of natural diet, and others are virulent in particular countries, arising from peculiarities. The Hindoos are considered the freest from disease of any part of the human race. The laborers on the African coast, who go from tribe to tribe to perform the manual labor, and whose strength is wonderful, live entirely on plain rice. The Irish, Swiss, and Gascons, the slaves of Europe, feed also on the simplest diet; the former chiefly on potatoes."

He states, also, that the diseases of cattle often afflict those who subsist on them. "In 1599," he observes, "the Venetian government, to stop a fatal disease among the people, prohibited the sale of meat, butter, or cheese, on Pain of death."

SIR ISAAC NEWTON.

This distinguished philosopher and mathematician is said to have abstained rigorously, at times, from all but purely vegetable food, and from all drinks but water; and it is also stated that some of his important labors were performed at these seasons of strict temperance. While writing his treatise on Optics, it is said he confined himself entirely to bread, with a little sack and water; and I have no doubt that his remarkable equanimity of temper, and that government of his animal appetites, throughout, for which he was so distinguished to the last hour of his life, were owing, in no small degree, to his habits of rigid temperance.

THE ABBE GALLANI.

The Abbé Gallani ascribes all social crimes to animal destruction—thus, treachery to angling and ensnaring, and murder to hunting and shooting. And he asserts that the man who would kill a sheep, an ox, or any unsuspecting animal, would, but for the law, kill his neighbor.

HOMER.

Even Homer, three thousand years ago, says Dr. Cheyne, could observe that the Homolgians—those Pythagoreans, those milk and vegetable eaters—were the longest lived and the honestest of men.

DR. BENJAMIN FRANKLIN.

Dr. Franklin, in his younger days, often, for some time together, lived exclusively on a vegetable diet, and that, too, in small quantity. During his after life he also observed seasons of abstinence from animal food, or *lents*, as he called them, of considerable length. His food and drink were, moreover, especially in early life, exceedingly simple; his meal often consisting of nothing but a biscuit and a slice of bread, with a bunch of raisins, and perhaps a basin of gruel. Now, Dr. F. testifies of himself; that he found his progress in science to be in proportion to that clearness of mind and aptitude of conception which can only be produced by total abstinence from animal food. He also derived many other advantages from his abstinence, both physical and moral.

MR. NEWTON.

This author wrote a work entitled "Defence of Vegetable Regimen." It is often quoted by Shelley, the poet, and others. I know nothing of the author or of his works, except through Shelley, who gives us some of his views, and informs us that seventeen persons, of all ages, consisting of Mr. Newton's family and the family of Dr. Lambe, who is elsewhere mentioned in this work, had, at the time he wrote, lived seven years on a pure vegetable diet, and without the slightest illness. Of the seventeen, some of them were infants, and one of them was almost dead with asthma when the experiment was commenced, but was already nearly cured by it; and of the family of Mr. N., Shelley testifies that they were "the most beautiful and healthy creatures it is possible to conceive"—the girls "perfect models for a sculptor"—and their dispositions "the most gentle and conciliating."

The following paragraph is extracted from Mr. Newton's "Defence," and will give us an idea of his sentiments. He was speaking of the fable of Prometheus:

"Making allowance for such transposition of the events of the allegory as time might produce after the important truths were forgotten, the drift of the fable seems to be this: Man, at his creation, was endowed with the gift of perpetual youth, that is, he was not formed to be a sickly, suffering creature, as we now see him, but to enjoy health, and to sink by slow degrees into the bosom of his parent earth, without disease or pain. Prometheus first taught the use of animal food, and of fire, with which to render it more digestible and pleasing to the taste. Jupiter and the rest of the gods, foreseeing the consequences of these inventions, were amused or irritated at the short-sighted devices of the newly-formed creature, and left him to experience the sad effects of them. Thirst, the necessary concomitant of a flesh diet, ensued; other drink than water was resorted to, and man forfeited the inestimable gift of health, which he had received from heaven; he became diseased, the partaker of a precarious existence, and no longer descended into his grave slowly."

O. S. FOWLER.

O. S. Fowler, the distinguished phrenologist, in his work on Physiology, devotes nearly one hundred pages to the discussion of the great diet question. He endeavors to show that, in every point of view, a flesh diet—or a diet partaking of flesh, fish, or fowl, in any degree—is inferior to a well-selected vegetable diet; and, as I think, successfully. He finally says:

"I wish my own children had never tasted, and would never taste, a mouthful of meat. Increased health, efficiency, talents, virtue, and happiness, would undoubtedly be the result. But for the fact that my table is set for others than my own wife and children, it would never be furnished with meat, so strong are my convictions against its utility."

I believe that L. N. Fowler, the brother and associate of the former, is of the same opinion; but my acquaintance with him is very limited. Both the Fowlers, with Mr. Wells, their associate in book-selling, seem anxiously engaged in circulating books which involve the discussion of this great question.

REV. MR. JOHNSTON.

Mr. Johnston, who for some fifteen or twenty years has been an American missionary in different foreign places—Trebizond, Smyrna, etc.—is, from conviction, a vegetable eater. The author holds in his possession several letters from this gentleman, on the subject of health, from which, but for want of room, he would be glad to make numerous extracts. He once sent, or caused to be sent, to him, at Trebizond, a barrel of choice American apples, for which the missionary, amid numerous Eastern luxuries, was almost starving. Happy would it be for many other American and British missionaries, if they had the same simple taste and natural appetite.

JOHN H. CHANDLER.

This young man has been for eight or ten years in the employ of the Baptist Foreign Missionary Board, and is located at Bangkok, in Siam. For several years before he left this country he was a vegetable eater, sometimes subsisting on mere fruit for one or two of his daily meals. And yet, as a mechanic, his labor was hard—sometimes severe.

Since he has been in Siam he has continued his reformed habits, as appears from his letters and from reports. The last letter I had from him was dated June 10, 1847. The following are extracts from it:

"I experienced the same trials (that is, from others) on my arrival in Burmah, in regard to vegetable diet, that I did in the United States. This I did not expect, and was not prepared for it. Through the blessing of God we were enabled to endure, and have persevered until now.

"Myself and wife are more deeply convinced than ever that vegetable diet is the best adapted to sustain health. I cannot say that we have been much more free from sickness than our associates; but one thing we can say—we have been equally well off, and our expenses have been much less."

After going on to say how much his family—himself and wife—saved by their plain living, viz., an average of about one dollar a week, he makes additional remarks, of which I will only quote the following:

"My labors, being mostly mechanical, are far more fatiguing than those of my brethren; and I do not think any of them could endure a greater amount of labor than I do."

It deserves to be noticed, in this connection, that Mr. Chandler has slender muscles, and would by no means be expected to accomplish as much as many men of greater vigor; and yet we have reason to believe that he performs as much labor as any man in the service of the board.

REV. JESSE CASWELL.

Mr. Caswell went out to India about thirteen years ago, a dyspeptic, and yet perhaps somewhat better than while engaged in his studies at Andover. For several years after his arrival he suffered much from sickness, like his fellow-laborers. His station was Bangkok. He was an American missionary, sent out by the American Board, as it is called, of Boston.

About six years ago he wrote me for information on the subject of health. He had read my works, and those of Mr. Graham, and seemed not only convinced of the general importance of studying the science of human life, but of the superiority of a well selected vegetable diet, especially at the East. He was also greatly anxious that missionaries should be early taught what he had himself learned. The following is one of his first paragraphs:

"I feel fully convinced that you are engaged in a work second to few if any of the great enterprises of the day. If there be any class of men standing in special need of correct physiological knowledge, that class consists of missionaries of the cross. What havoc has disease made with this class, and for the most part, as I feel convinced, because, before and after leaving their native land, they live so utterly at variance with the laws of their nature."

He then proceeds to say, that the American missionaries copy the example of the English, and that they all eat too much high-seasoned food, and too much flesh and fish; and argues against the practice by adducing facts. The following is one of them:

"My Siamese teacher, a man about forty years old, says that those who live simply on rice, with a little salt, enjoy better health, and can endure a greater amount of labor, than those who live in any other way. * * * The great body of the Siamese use no flesh, except fish. Of this they generally eat *a very little*, with their rice."

The next year I had another letter from him. He had been sick, but was better, and thought he had learned a great deal, during his sickness, about the best means of preserving health. He had now fully adopted what he chose to call the Graham system, and was rejoicing—he and his wife and children—in its benefits. He says, "If a voice from an obscure corner of the earth can do any thing toward encouraging your heart and staying your hands, that voice you shall have." He suggests the propriety of my sending him a copy of "Vegetable Diet." "I think," says he, "it might do great good." He wished to lend it among his friends.

It must suffice to say, that he continued to write me, once or twice a year, as long as he lived. He also insisted strongly on the importance of physiological information among students preparing for the ministry, and especially for missions. He even wrote once or twice to Rev. Dr. Anderson, and solicited attention to the subject. But the board would neither hear to him nor to me, except to speak kind words, for nothing effective was ever done. They even refused a well-written communication on the subject, intended for the Missionary Herald. Let me also say, that as early as March, 1845, he told me that Dr. Bradley, his associate (now in this country), with his family, were beginning to live on the vegetable system; and added, that one of the sisters of the mission, who was no "Grahamite," had told him she thought there was not one third as much flesh used in all the mission families that there was a year before.

Mr. Caswell became exceedingly efficient, over-exerted himself in completing a vocabulary of the Siamese language, and in other labors, and died in September last. He was, according to the testimony of Dr. Bradley, a "*noble man*;" and probably his life and health, and that of his family, were prolonged many years by his improved habits. But his early transgressions— like those of thousands—at length found him out. I allude to his errors in regard to exercise, eating, drinking, sleeping, taking medicine, etc.

MR. SAMUEL CHINN.

This individual has represented the town of Marblehead, Mass., in the state legislature, and is a man of respectability. He is now, says the "Lynn Washingtonian," above forty years of age, a strong, healthy man, and, to use his own language, "has neither ache nor pain." For the ten years next preceding our last account from him he had lived on a simple vegetable diet, condemning to slaughter no flocks or herds that "range the valley free," but leaving them to their native, joyous hill-sides and mountains. But Mr. Chinn, not contented with abstinence from animal food, goes nearly the full length of Dr. Schlemmer and his sect, and abjures cookery. For four years he subsisted—we believe he does so now—on nothing but unground wheat and fruit. His breakfast, it is said, he uniformly makes of fruit; his other two meals of unground wheat; patronizing neither millers nor cooks. A few years since, being appointed a delegate to a convention in Worcester, fifty-eight miles distant, he filled his pocket with wheat, walked there during the day, attended the convention, and the next day walked home again, with comparative ease.

FATHER SEWALL.

This venerable man—Jotham Sewall, of Maine, as he styles himself, one of the fathers of that state—is now about ninety years of age, and yet is, what he has long been, an active home missionary. He is a man of giant size and venerable appearance, of a green old age, and remarkably healthy. He is an early riser, a man of great cheerfulness, and of the most simple habits. He has abstained from tea and coffee—poisonous things, as he calls them—forty-seven years. His only drinks are water and sage tea. These, with bread, milk, and fruits, and perhaps a little salt, are the only things that enter his stomach. How long he has abstained from flesh and fish I have not learned, but I believe some thirty or forty years.

Such is the appearance of this venerable man, that no one is surprised to find in him those gigantic powers of mind, and that readiness to give wise counsel on every important occasion, for which he has so long been distinguished. It has sometimes seemed to me that no one would doubt the efficacy of a well-selected vegetable diet to give strength, mental or bodily, who had known Father Sewall.

MAGLIABECCHI,

An Italian, who died in the beginning of the eighteenth century, abjured cookery at the age of forty years, and confined himself chiefly to fruits, grains, and water. He never allowed himself a bed, but slept on a kind of settee, wrapped in a long morning gown, which served him for blanket and clothing the year round.

I would not be understood as encouraging the anti-cookery system of Dr. Schlemmer and Magliabecchi; but it is interesting to know *what can be done*. Magliabecchi lived to the age of from eighty to one hundred years.

OBERLIN AND SWARTZ.

These two distinguished men were essentially vegetable eaters. Of the habits of Oberlin, the venerable pastor and father of Waldbach, I am not able to speak, however, with so much certainty as of those of Swartz. His income, during the early part of his residence in India, was only forty-eight pounds a year, which, being estimated by its ability to procure supplies for his necessities, was only equal to about one hundred dollars. He not only accepted of very narrow quarters, but ate, drank, and dressed, in the plainest manner. "A dish of rice and vegetables," says his biographer, "satisfied his appetite for food."

THE IRISH.

Much has been said of the dietetic habits of the Irish, of late years, especially of their potato. Now, we have abundant facts which go to prove that good potatoes form a wholesome aliment, equal, if not superior, to many forms of European and American diet. Yet it cannot be that a diet consisting wholly of potatoes is as well for the race as one partaking of greater variety.

Mr. Gamble, a traveler in Ireland, in his work on Irish "Society and Manners," gives the following statement of an old friend of his, whom he visited:

"He was upward of eighty years when I had last seen him, and he was now in his ninety-fourth year. He found the old gentleman seated on a kind of rustic seat, in the garden, by the side of some bee-hives. He was asleep. On his waking I was astonished to see the little change time had wrought on him; a little more stoop in his shoulders, a wrinkle more, perhaps, in his forehead, a more perfect whiteness of his hair, was all the difference since I had seen him last. Flesh meat in my venerable friend's house was an article never to be met with. *For sixty years past he had not tasted it*, nor did he by any means like to see it taken by others. His food was vegetables, bread, milk, butter, and honey. His whole life was a series of benevolent actions, and Providence rewarded him, even here, by a peace of mind which passeth all understanding, by a judgment vigorous and unclouded, and by a length of days beyond the common course of men."

James Haughton, I believe of Dublin—a correspondent of Henry C. Wright, of Philadelphia, who is himself in theory a vegetable eater—has, for some time past, rejected flesh, and pursued a simple course of living, as he says, with great advantage. I have been both amused and instructed by his letters.

I have met with several Irish people of intelligence who were vegetable eaters, but their names are not now recollected. They have not, however, in any instance, confined themselves to potatoes. One of the most distinguished of these was a female laborer in the family of a merchant at Barnstable. She was, from choice, a very rigid vegetable eater; and yet no person in the whole neighborhood was more efficient as a laborer. Those who know her, and are in the habit of thinking no person can work hard without flesh and fish, often express their astonishment that she should be able to live so simply and yet perform so much labor.

JOHN BAILIES.

John Bailies, of England, who reached the great age of one hundred and twenty-eight, is said to have been a strict vegetarian. His food, for the most part, consisted of brown bread and cheese; and his drink of water and milk. He had survived the whole town of Northampton (as he was wont to say), where he resided, three or four times over; and it was his custom to say that they were all killed by tea and coffee. Flesh meat at that time had not come into suspicion, otherwise he would doubtless have attributed part of the evil to this agency.

FRANCIS HUPAZOLI.

This gentleman was a Sardinian ecclesiastic, at the first; afterward a merchant at Scio; and finally Venetian consul at Smyrna. Much has been said of Lewis Cornaro, who, having broken down his constitution at the age of forty, renewed it by his temperance, and lasted unto nearly the age of a century. His story is interesting and instructive; but little more so than that of Hupazoli.

His habits were all remarkable for simplicity and truth, except one. He was greatly licentious; and his licentiousness, at the age of eighty-five, had nearly carried him off. Yet such was the mildness of his temper, and so correct was he in regard to exercise, rest, rising, eating, drinking, etc., that he lived on, to the great age of one hundred and fifteen years, and then died, not of old age, but of disease.

Hupazoli did not entirely abstain from flesh; and yet he used very little, and that was wild game. His living was chiefly on fruits. Indeed, he ate but little at any time; and his supper was particularly light. His drink was water. He never took any medicine in his whole life, not even tobacco; nor was he so much as ever bled. In fact, till late in life, he was never sick.

MARY CAROLINE HINCKLEY.

This young woman, a resident of Hallowell, in Maine, and somewhat distinguished as a poet, is, from her own conviction and choice both, a vegetable eater. Her story, which I had from her friends, is substantially as follows:

When about eleven years of age she suddenly changed her habits of eating, and steadfastly refused, at the table, all kinds of food which partook of flesh and fish. The family were alarmed, and afraid she was ill. When they made inquiry concerning it, she hesitated to assign the reasons for her conduct; but, on being pressed closely, she confessed that she abstained for conscience' sake; that she had become fully convinced, from reading and reflection, that she ought not to eat animal food.

It was in vain that the family and neighbors remonstrated with her, and endeavored, in various ways, to induce her to vary from her purpose. She continued to use no fowl, flesh, or fish; and in this habit she continues, as I believe, to this day, a period of some twelve or fifteen years.

JOHN WHITCOMB.

John Whitcomb, of Swansey, N. H., at the age of one hundred and four was in possession of sound mind and memory, and had a fresh countenance; and so good was his health, that he rose and bathed himself in cold water even in mid-winter. His wounds, moreover, would heal like those of a child. And yet this man, for eighty years, refused to drink any thing but water; and for thirty years, at the close of life, confined himself chiefly to bread and milk as his diet.

CAPT. ROSS, OF THE BRITISH NAVY.

It is sometimes said that animal food is indispensably necessary in the polar regions. We have seen, however, in the testimony of Professor Sweetser, that this view of the case is hardly correct. But we have positive testimony on this subject from Capt. Ross himself.

This navigator, with his company, spent the winter of 1830-31 above 70° of north latitude, without beds, clothing (that is, extra clothing), or animal food, and with no evidence of any suffering from the mere disuse of flesh and fish.

HENRY FRANCISCO.

This individual, who died at Whitehall, N. Y., in the year 1820, at the age of one hundred and twenty-five years, was, during the latter part of his life, quite a Grahamite, as the moderns would call him. His favorite articles of food were tea, bread and butter, and baked apples; and he was even abstemious in the use of these.

PROFESSOR FERGUSON.

Professor Adam Ferguson, an individual not unknown in the literary world, was, till he was fifty years of age, regarded as quite healthy. Brought up in fashionable society, he was very often invited to fashionable dinners and parties, at which he ate heartily and drank wine—sometimes several bottles. Indeed, he habitually ate and drank freely; and, as he had by nature a very strong constitution, he thought nothing which he ate or drank injured him.

Things went on in this manner, as I have already intimated, till he was fifty years of age. One day, about this time, having made a long journey in the cold, he returned very much fatigued, and in this condition went to dine with a party, where he ate and drank in his usual manner. Soon after dinner, he was seized with a fit of apoplexy, followed by palsy; but by bleeding, and other energetic measures, he was partially restored.

He was now, by the direction of his physician, put upon what was called a low diet. It consisted of vegetable food and milk. For nearly forty years he tasted no meat, drank nothing but water and a little weak tea, and took no suppers. If he ventured, at any time, upon more stimulating food or drink, he soon had a full pulse, and hot, restless nights. His bowels, however, seemed to be much affected by the fit of palsy; and not being inclined, so far as I can learn, to the use of fruit and coarse bread, he was sometimes compelled to use laxatives.

When he was about seventy years of age, however, all his paralytic symptoms had disappeared; and his health was so excellent, for a person of his years, as to excite universal admiration. This continued till he was nearly ninety. His mind, up to this time, was almost as entire as in his younger days; none of his bodily functions, except his sight, were much impaired. So perfect, indeed, was the condition of his physical frame, that nobody, who had not known his history, would have suspected he had ever been apoplectic or paralytic.

When about ninety years of age, his health began slightly to decline. A little before his death, he began to take a little meat. This, however, did not save him—nature being fairly worn out. On the contrary, it probably hastened his dissolution. His bowels became irregular, his pulse increased, and he fell into a bilious fever, of which he died at the great age of ninety-three.

Probably there are, on record, few cases of longevity more instructive than this. Besides showing the evil tendency of living at the expense of life, it also shows, in a most striking manner, the effects of simple and unstimulating food and drink, even in old age; and the danger of recurring to the use of that which is more stimulating in very advanced life. In this last respect, it confirms the experience of Cornaro, who was made sick by attempting, in his old age, and at the solicitation of kind friends, to return to the use of a more stimulating diet; and of Parr, who was destroyed in the same way, after having attained to more than a hundred and fifty years.

But the fact that living at the expense of life, cuts down, here and there, in the prime of life, or even at the age of fifty, a few individuals, though this of itself is no trivial evil, is not all. Half of what we call the infirmities of old age—and thus charge them upon Him who made the human frame *subject* to age—have their origin in the same source; I mean in this living too fast, and exhausting prematurely the vital powers. When will the sons of men learn wisdom in this matter? Never, I fear, till they are taught, as commonly as they now are reading and writing, the principles of physiology.

HOWARD, THE PHILANTHROPIST.

Although individual cases of abstinence from animal food prove but little, yet they prove something in the case of a man so remarkable as John Howard. If he, with a constitution not very strong, and in the midst of the greatest fatigues of body and mind, could best sustain himself on a bread and water, or bread and tea diet, who is there that would not be well sustained on vegetable food? And yet it is certain that Howard was a vegetable eater for many years of the latter part of his life; and that had he not exposed himself in a remarkable manner, there is no known reason why he might not have lasted with a constitution no better than his was, to a hundred years of age.

GEN. ELLIOTT.

The following extract exhibits in few words, the dietetic history of that brave and wise commander, General George Augustus Elliott, of the British army:

"During the whole of his active life, Gen. Elliott had inured himself to the most rigid habits of order and watchfulness; seldom sleeping more than four hours a day, and never eating any thing but vegetable food, or drinking any thing but water. During eight of the most anxious days of the memorable siege of Gibraltar, he confined himself to four ounces of rice a day. He was universally regarded as one of the most abstemious men of his age.

"And yet his abstemiousness did not diminish his vigor; for, at the abovementioned siege of Gibraltar, when he was sixty-six years of age, he had nearly all the activity and fire of his youth. Nor did he die of any wasting disease, such as full feeders are wont to say men bring upon them by their abstinence. On the contrary, owing to a hereditary tendency, perhaps, of his family, he died at the age of seventy-three, of apoplexy."

ENCYCLOPEDIA AMERICANA.

The following testimony is from the Encyclopedia. I do not suppose the writer was the friend of a diet exclusively vegetable; but his testimony is therefore the more interesting. His only serious mistake is in regard to the tendency of vegetable food to form weak fibres.

"Sometimes a particular kind of food is called wholesome, because it produces a beneficial effect of a particular character on the system of an individual. In this case, however, it is to be considered as a medicine; and can be called wholesome only for those whose systems are in the same condition.

"Aliments abounding in fat are unwholesome, because fat resists the operation of the gastric juice.

"The addition of too much spice makes many an innocent aliment injurious, because spices resist the action of the digestive organs, and produce an irritation of particular parts of the system.

"The kind of aliment influences the health, and even the character of man. He is fitted to derive nourishment both from animal and vegetable aliment; but can live exclusively on either.

"Experience proves that animal food most readily augments the solid parts of the blood, the fibrine, and therefore the strength of the muscular system; but disposes the body, at the same time, to inflammatory, putrid, and scorbutic diseases; and the character to violence and coarseness. On the contrary, vegetable food renders the blood lighter and more liquid, but forms weak fibres, disposes the system to the diseases which spring from feebleness, and tends to produce a gentle character.

"Something of the same difference of moral effect results from the use of strong or light wines. But the reader must not infer that meat is indispensable for the support of the bodily strength. The peasants of some parts of Switzerland, who hardly ever taste any thing but bread, cheese, and butter, are vigorous people.

"The nations of the north are inclined, generally, more to animal aliment; those of the south and the Orientals, more vegetable. The latter are generally more simple in their diet than the former, when their taste has not been corrupted by luxurious indulgence. Some tribes in the East, and the caste of Bramins in India, live entirely on vegetable food."

MR. THOMAS BELL, OF LONDON.

Mr. Thomas Bell, Fellow of the Royal Society, Member of the Royal College of Surgeons in London, Lecturer on the Anatomy and Diseases of the Teeth, at Guy's Hospital, and Surgeon Dentist to that institution, in his physiological observations on the natural food of man, deduced from the character of the teeth, says, "The opinion which I venture to give, has not been hastily formed, nor without what appeared to me sufficient grounds. It is not, I think, going too far to say, that every fact connected with human organization goes to prove that man was originally formed a frugiverous (fruit-eating) animal, and therefore, probably, tropical or nearly so, with regard to his geographical situation. This opinion is principally derived from the formation of his teeth and digestive organs, as well as from the character of his skin and general structure of his limbs."

LINNÆUS, THE NATURALIST.

Linnæus, in speaking of fruits and esculent vegetables, says—"This species of food is that which is most suitable to man, as is evinced by the structure of the mouth, of the stomach, and of the hands."

SHELLEY, THE POET.

The following are the views of that eccentric, though in many respects sensible writer, Shelley, as presented in a note to his work, called Queen Mab. I have somewhat abridged them, not solely to escape part of his monstrous religious sentiments, but for other reasons. I have endeavored, however, to preserve, undisturbed, his opinions and reasonings, which I hope will make a deep and abiding impression:

"The depravity of the physical and moral nature of man, originated in his unnatural habits of life. The language spoken by the mythology of nearly all religions seems to prove that, at some distant period, man forsook the path of nature, and sacrificed the purity and happiness of his being to unnatural appetites. Milton makes Raphael thus exhibit to Adam the consequence of his disobedience:

'——Immediately, a placeBefore his eyes appeared; and, noisome, dark,A lazar-house it seemed; wherein were laidNumbers of all diseased; all maladiesOf ghastly spasm, or racking torture, qualmsOf heart-sick agony, all feverous kinds,Convulsions, epilepsies, fierce catarrhs,Intestine stone and ulcer, colic pangs,Demoniac frenzy, moping melancholy,And moon-struck madness, pining atrophy,Marasmus, and wide-wasting pestilence,Dropsies and asthmas, and joint-racking rheums.'

"The fable of Prometheus, too, is explained in a manner somewhat similar. Before the time of Prometheus, according to Hesiod, mankind were exempt from suffering; they enjoyed a vigorous youth; and death, when at length it came, approached like sleep, and gently closed the eyes. Prometheus (who represents the human race) effected some great change in the condition of his nature, and applied fire to culinary purposes. From this moment his vitals were devoured by the vulture of disease. It consumed his being in every shape of its loathsome and infinite variety, inducing the soul-quelling sinkings of premature and violent death. All vice arose from the ruin of healthful innocence.

"Man, and the animals which he has infected with his society, or depraved by his dominion, are alone diseased. The wild hog, the bison, and the wolf are perfectly exempt from malady, and invariably die, either from external violence or natural old age. But the domestic hog, the sheep, the cow, and the dog are subject to an incredible number of distempers, and, like the corrupters of their nature, have physicians, who thrive upon their miseries.

"The supereminence of man is like Satan's supereminence of pain,—and the majority of his species, doomed to penury, disease, and crime, have reason to curse the untoward event, that, by enabling him to communicate his sensations, raised him above the level of his fellow animals. But the steps that have been taken are irrevocable.

"The whole of human science is comprised in one question: How can the advantages of intellect and civilization be reconciled with the liberty and pure pleasures of natural life? How can we take the benefits and reject the evils of the system, which is now interwoven with our being? I believe that *abstinence from animal food and spirituous liquors would, in a great measure, capacitate us for the solution of this important question.*

"It is true, that mental and bodily derangement is attributable in part to other deviations from rectitude and nature than those which concern diet. The mistakes cherished by society respecting the connection of the sexes, whence the misery and diseases of celibacy, unenjoying prostitution, and the premature arrival of puberty, necessarily spring; the putrid atmosphere of crowded cities; the exhalations of chemical processes: the muffling of our bodies in superfluous apparel; the absurd treatment of infants; all these, and innumerable other causes, contribute their mite to the mass of human evil.

"Comparative anatomy teaches us that man resembles frugiverous animals in every thing, and carnivorous in nothing; he has neither claws wherewith to seize his prey, nor distinct and pointed teeth to tear the living fibre. A mandarin of the first class, with nails two inches long, would probably find them, alone, inefficient to hold even a hare. It is only by softening and disguising dead flesh by culinary preparations that it is rendered susceptible of mastication and digestion, and that the sight of its bloody juices does not excite intolerable loathing, horror, and disgust. Let the advocate of animal food force himself to a decisive experiment on its fitness, and, as Plutarch recommends, tear a living lamb with his teeth, and, plunging his head into its vitals, slake his thirst with the steaming blood; when fresh from the deed of horror, let him revert to the irresistible instincts of nature that would rise in judgment against it, and say, Nature formed me for such work as this. Then, and then only, would he be consistent.

"Young children evidently prefer pastry, oranges, apples, and other fruit, to the flesh of animals, until, by the gradual depravation of the digestive organs, the free use of vegetables has, for a time, produced serious inconveniences. *For a time*, I say, since there never was an instance wherein a change from spirituous liquors and animal food to vegetables and pure water has failed ultimately to invigorate the body, by rendering its juices bland and consentaneous, and to restore to the mind that cheerfulness and elasticity which not one in fifty possesses on the present system. A love of strong liquor is also with difficulty taught to infants. Almost every one remembers the wry faces which the first glass of port produced. Unsophisticated instinct is invariably unerring; but to decide on the fitness of animal food from the perverted appetites which its constrained adoption produces, is to make the criminal a judge in his own cause; it is even worse—it is appealing to the infatuated drunkard in a question of the salubrity of brandy.

"Except in children, however, there remain no traces of that instinct which determines, in all other animals, what aliment is natural or otherwise; and so perfectly obliterated are they in the reasoning adults of our species, that it has become necessary to urge considerations drawn from comparative anatomy to prove that we are naturally frugiverous.

"Crime is madness. Madness is disease. Whenever the cause of disease shall be discovered, the root, from which all vice and misery have so long overshadowed the globe, will be bare to the axe. All the exertions of man, from that moment, may be considered as tending to the clear profit of his species. No sane mind, in a sane body, resolves upon a crime. It is a man of violent passions, blood-shot eyes, and swollen veins, that alone can grasp the knife of murder. The system of a simple diet is not a reform of legislation, while the furious passions and evil propensities of the human heart, in which it had its origin, are unassuaged. It strikes at the root of all evil, and is an experiment which may be tried with success, not alone by nations, but by small societies, families, and even individuals. In no case has a return to a vegetable diet produced the slightest injury; in most it has been attended with changes undeniably beneficial.

"Should ever a physician be born with the genius of Locke, he might trace all bodily and mental derangements to our unnatural habits, as clearly as that philosopher has traced all knowledge to sensation. What prolific sources of disease are not those mineral and vegetable poisons, that have been introduced for its extirpation! How many thousands have become murderers and robbers, bigots and domestic tyrants, dissolute and abandoned adventurers, from the use of fermented liquors, who, had they slaked their thirst only with pure water, would have lived but to diffuse the happiness of their own unperverted feelings! How many groundless opinions and absurd institutions have not received a general sanction from the sottishness and intemperance of individuals!

"Who will assert that, had the populace of Paris satisfied their hunger at the ever-furnished table of vegetable nature, they would have lent their brutal suffrage to the proscription-list of Robespierre? Could a set of men, whose passions were not perverted by unnatural stimuli, look with coolness on an *auto da fé*? Is it to be believed that a being of gentle feelings, rising from his meal of roots, would take delight in sports of blood?

"Was Nero a man of temperate life? Could you read calm health in his cheek, flushed with ungovernable propensities of hatred for the human race? Did Muley Ismail's pulse beat evenly? was his skin transparent? did his eyes beam with healthfulness, and its invariable concomitants, cheerfulness and benignity?

"Though history has decided none of these questions, a child could not hesitate to answer in the negative. Surely the bile-suffused cheek of Bonaparte, his wrinkled brow, and yellow eye, the ceaseless inquietude of his nervous system, speak no less plainly the character of his unresting ambition than his murders and his victories. It is impossible, had Bonaparte descended from a race of vegetable feeders, that he could have had either the inclination or the power to ascend the throne of the Bourbons.

"The desire of tyranny could scarcely be excited in the individual; the power to tyrannize would certainly not be delegated by a society neither frenzied by inebriation nor rendered impotent and irrational by disease. Pregnant, indeed, with inexhaustible calamity is the renunciation of instinct, as it concerns our physical nature. Arithmetic cannot enumerate, nor reason perhaps suspect, the multitudinous sources of disease in civilized life. Even common water, that apparently innoxious *pabulum*, when corrupted by the filth of populous cities, is a deadly and insidious destroyer.

"There is no disease, bodily or mental, which adoption of vegetable diet and pure water has not infallibly mitigated, wherever the experiment has been fairly tried. Debility is gradually converted into strength, disease into healthfulness; madness, in all its hideous variety, from the ravings of the fettered maniac, to the unaccountable irrationalities of ill-temper, that make a hell of domestic life, into a calm and considerate evenness of temper, that alone might offer a certain pledge of the future moral reformation of society.

"On a natural system of diet, old age would be our last and our only malady; the term of our existence would be protracted; we should enjoy life, and no longer preclude others from the enjoyment of it; all sensational delights would be infinitely more exquisite and perfect; the very sense of being would then be a continued pleasure, such as we now feel it in some few and favored moments of our youth.

"By all that is sacred in our hopes for the human race, I conjure those who love happiness and truth, to give a fair trial to the vegetable system. Reasoning is surely superfluous on a subject whose merits an experience of six months should set forever at rest.

"But it is only among the enlightened and benevolent that so great a sacrifice of appetite and prejudice can be expected, even though its ultimate excellence should not admit of dispute. It is found easier by the short-sighted victims of disease, to palliate their torments, by medicine, than to prevent them by regimen. The vulgar of all ranks are invariably sensual and indocile; yet I cannot but feel myself persuaded, that when the benefits of vegetable diet are mathematically proved—when it is as clear, that those who live naturally are exempt from premature death, as that nine is not one, the most sottish of mankind will feel a preference toward a long and tranquil, contrasted with a short and painful life.

"On the average, out of sixty persons, four die in three years. Hopes are entertained, that in April, 1814,[20] a statement will be given that sixty persons, all having lived more than three years on vegetables and pure water, are then in *perfect health*. More than two years have now elapsed; *not one of them has died*; no such example will be found in any sixty persons taken at random.

"When these proofs come fairly before the world, and are clearly seen by all who understand arithmetic, it is scarcely possible that abstinence from aliments demonstrably pernicious should not become universal.

"In proportion to the number of proselytes, so will be the weight of evidence; and when a thousand persons can be produced, living on vegetables and distilled water, who have to dread no disease but old age, the world will be compelled to regard animal flesh and fermented liquors as slow but certain poisons.

"The change which would be produced by simple habits on political economy, is sufficiently remarkable. The monopolizing eater of animal flesh would no longer destroy his constitution by devouring an acre at a meal, and many loaves of bread would cease to contribute to gout, madness, and apoplexy, in the shape of a pint of porter, or a dram of gin, when appeasing the long-protracted famine of the hard-working peasant's hungry babes.

"The quantity of nutritious vegetable matter, consumed in fattening the carcass of an ox, would afford ten times the sustenance, undepraving indeed, and incapable of generating disease, if gathered immediately from the bosom of the earth. The most fertile districts of the habitable globe are now actually cultivated by men for animals, at a delay and waste of aliment absolutely incapable of calculation. It is only the wealthy that can, to any great degree, even now, indulge the unnatural craving for dead flesh, and they pay for the greater license of the privilege, by subjection to supernumerary diseases.

"Again—the spirit of the nation that should take the lead in this great reform would insensibly become agricultural; commerce, with its vices, selfishness, and corruption, would gradually decline; more natural habits would produce gentler manners, and the excessive complication of political relations would be so far simplified that every individual might feel and understand why he loved his country, and took a personal interest in its welfare.

"On a natural system of diet, we should require no spices from India; no wines from Portugal, Spain, France, or Madeira; none of those multitudinous articles of luxury, for which every corner of the globe is rifled, and which are the cause of so much individual rivalship, and such calamitous and sanguinary national disputes.

"Let it ever be remembered, that it is the direct influence of excess of commerce to make the interval between the rich and the poor wider and more unconquerable. Let it be remembered, that it is a foe to every thing of real worth and excellence in the human character. The odious and disgusting aristocracy of wealth, is built upon the ruins of all that is good in chivalry or republicanism; and luxury is the forerunner of a barbarism scarce capable of cure. Is it impossible to realize a state of society, where all the energies of man shall be directed to the production of his solid happiness?

"None must be intrusted with power (and money is the completest species of power), who do not stand pledged to use it exclusively for the general benefit. But the use of animal flesh and fermented liquors, directly militates with this equality of the rights of man. The peasant cannot gratify these fashionable cravings without leaving his family to starve. Without disease and war, those sweeping curtailers of population, pasturage would include a waste too great to be afforded. The labor requisite to support a family is far lighter than is usually supposed. The peasantry work, not only for themselves, but for the aristocracy, the army, and the manufacturers.

"The advantage of a reform in diet is obviously greater than that of any other. It strikes at the root of the evil. To remedy the abuses of legislation, before we annihilate the propensities by which they are produced, is to suppose that by taking away the effect, the cause will cease to operate.

"But the efficacy of this system depends entirely on the proselytism of individuals, and grounds its merits, as a benefit to the community, upon the total change of the dietetic habits in its members. It proceeds securely from a number of particular cases to one that is universal, and has this advantage over the contrary mode, that one error does not invalidate all that has gone before.

"Let not too much, however, be expected from this system. The healthiest among us is not exempt from hereditary disease. The most symmetrical, athletic, and long-lived is a being inexpressibly inferior to what he would have been had not the unnatural habits of his ancestors accumulated for him a certain portion of malady and deformity. In the most perfect specimen of civilized man, something is still found wanting by the physiological critic. Can a return to nature, then, instantaneously eradicate predispositions that have been slowly taking root in the silence of innumerable ages? Indubitably not. All that I contend for is, that from the moment of relinquishing all unnatural habits, no new disease is generated; and that the predisposition to hereditary maladies gradually perishes for want of its accustomed supply. In cases of consumption, cancer, gout, asthma, and scrofula, such is the invariable tendency of a diet of vegetables and pure water.

"Those who may be induced by these remarks to give the vegetable system a fair trial, should, in the first place, date the commencement of their practice from the moment of their conviction. All depends upon breaking through a pernicious habit resolutely and at once. Dr. Trotter asserts, that no drunkard was ever reformed by gradually relinquishing his dram. Animal flesh, in its effects on the human stomach, is analogous to a dram; it is similar to the kind, though differing in the degree of its operation. The proselyte to a pure diet must be warned to expect a temporary diminution of muscular strength. The subtraction of a powerful stimulus will suffice to account for this event. But it is only temporary, and is succeeded by an equable capability for exertion, far surpassing his former various and fluctuating strength.

"Above all, he will acquire an easiness of breathing, by which such exertion is performed, with a remarkable exemption from that painful and difficult panting now felt by almost every one, after hastily climbing an ordinary mountain. He will be equally capable of bodily exertion or mental application, after, as before his simple meal. He will feel none of the narcotic effects of ordinary diet. Irritability, the direct consequence of exhausting stimuli, would yield to the power of natural and tranquil impulses. He will no longer pine under the lethargy of *ennui*, that unconquerable weariness of life, more to be dreaded than death itself.

"He will no longer be incessantly occupied in blunting and destroying those organs from which he expects his gratification. The pleasures of taste to be derived from a dinner of potatoes, beans, peas, turnips, lettuce, with a dessert of apples, gooseberries, strawberries, currants, raspberries, and in winter, oranges, apples, and pears, is far greater than is supposed. Those who wait until they can eat this plain fare with the sauce of appetite, will scarcely join with the hypocritical sensualist at a lord mayor's feast, who declaims against the pleasures of the table."

REV. EZEKIEL RICH.

This gentleman, once a teacher in Troy, N. H., now nearly seventy years of age, is a giant, both intellectually and physically, like Father Sewall, of Maine. The following is his testimony—speaking of what he calls his system:

"Such a system of living was formed by myself, irrespective of Graham or Alcott, or any other modern dietetic philosophers and reformers, although I agree with them in many things. It allows but little use of flesh, condiments, concentrated articles, complex cooking, or hot and stimulating drinks. On the other hand, it requires great use of milk, the different bread stuffs, fruits, esculent roots and pulse, all well, simply, and neatly cooked."

REV. JOHN WESLEY.

The habits of this distinguished individual, though often adverted to, are yet not sufficiently known. For the last half of his long life (eighty-eight years) he was a thorough going vegetarian. He also testifies that for three or four successive years he lived entirely on potatoes; and during that whole time he never relaxed his arduous ministerial labors, nor ever enjoyed better health.

LAMARTINE.

Lamartine was educated a vegetarian of the strictest sort—an education which certainly did not prevent his possessing as fine a physical frame as can be found in the French republic. Of his mental and moral characteristics it is needless that I should speak. True it is that Lamartine ate flesh and fish at one period of his life; but we have the authority of Douglas Jerrold's London Journal for assuring our readers that he is again a vegetarian.

FOOTNOTES:

[17] Some, however, represent the great apostle to have been a rigid vegetable eater. On this point I have no settled opinion.

[18] It may be found at full length at page 233 of the 6th volume of the Library of Health.

[19] Instances, he says, are not rare (but this I doubt), of two hundred children born to a man by his different wives, in some parts of the interior of Africa.

[20] A date but little later than that of the work whence this article is extracted.

CHAPTER VII.

SOCIETIES AND COMMUNITIES ON THE VEGETABLE SYSTEM.

The Pythagoreans.—The Essenes.—The Bramins.—Society of Bible Christians.—Orphan Asylum of Albany.—The Mexican Indians.—School in Germany.—American Physiological Society.

GENERAL REMARKS.

The following chapter did not come within the scope of my plan, as it was originally formed. But in prosecuting the labors of preparing a volume on vegetable diet, it has more and more seemed to me desirable to add a short account of some of the communities and associations of men, both of ancient and modern times, who, amid a surrounding horde of flesh-eaters, have withstood the power of temptation, and proved, in some measure, true to their own nature, and the first impulses of mercy, humanity, and charity. I shall not, of course, attempt to describe all the sects and societies of the kind to which I refer, but only a few of those which seem to me most important.

One word may be necessary in explanation of the term communities. I mean by it, smaller communities, or associations. There have been, and still are, many whole nations which might be called vegetable-eating communities; but of such it is not my purpose to speak at present.

THE PYTHAGOREANS.

Pythagoras appears to have flourished about 550 years before Christ. He was, probably, a native of the island of Samos; but a part of his education, which was extensive and thorough, was received in Egypt. He taught a new philosophy; and, according to some, endeavored to enforce it by laying claim to supernatural powers. But, be this as it may have been, he was certainly a man of extraordinary qualities and powers, as well as of great and commanding influence. In an age of great luxury and licentiousness, he taught, both by example and precept, the most rigid doctrines of sobriety, temperance, and purity. He abstained from all animal food, and limited himself entirely to vegetables; of which he usually preferred bread and honey. Nor did he allow the free use of every kind of vegetable; for beans, and I believe every species of pulse, were omitted. Water was his only drink. He lived, it is said, to the age of eighty; and even then did not perish from disease or old age, but from starvation in a place where he had sought a retreat from the fury of his enemies.

His disciples are said to have been exceedingly numerous, in almost all quarters of the then known world, especially in Greece and Italy. It is impossible, however, to form any conjecture of their numbers. The largest school or association of his rigid followers is supposed to have been at the city of Crotona, in South Italy. Their number was six hundred. They followed all his dietetic and philosophical rules with the utmost strictness. The association appears to have been, for a time, exceedingly flourishing. It was a society of philosophers, rather than of common citizens. They held their property in one common stock, for the benefit of the whole. The object of the association was chiefly to aid each other in promoting intellectual cultivation. Pythagoras did not teach abstinence from all hurtful food and drink, and an exclusive use of that which was the *best*, for the sole purpose of making men better, or more healthy, or longer-lived *animals*; he had a higher and nobler purpose. It was to make them better rationals, more truly noble and god-like—worthy the name of rational men, and of the relation in which they stood to their common Father. And yet, after all, his doctrines appear to have been mingled with much bigotry and superstition.

THE ESSENES.

The following account of this singular sect of the ancient Jews is abridged from an article in the Annals of Education, for July, 1836. The number of this vegetable-eating sect is not known, though, according to Philo, there were four thousand of them in the single province of Judea.

"Pliny, says that the Essenes of Judea fed on the fruit of the palm-tree. But, however this may have been, it is agreed, on all hands, that, like the ancient Pythagoreans, they lived exclusively on vegetable food, and that they were abstinent in regard to the quantity even of this. They would not kill a living creature, even for sacrifices. It is also understood that they treated diseases of every kind—though it does not appear that they were subject to many—with roots and herbs. Josephus says they were long-lived, and that many of them lived over a hundred years. This he attributes to their 'regular course of life,' and especially to 'the simplicity of their diet.'"

THE BRAMINS.

The Bramins, or Brahmins, are, as is probably well known, the first of the four *castes* among the Hindoos. They are the priests of the people, and are remarkable, in their way, for their sanctity. Of their number I am not at present apprised, but it must be very great. But, however great it may be, they are vegetable eaters of the strictest sect. They are not even allowed to eat eggs; and I believe milk and its products are also forbidden them; but of this I am not quite certain. Besides adhering to the strictest rules of temperance, they are also required to observe frequent fasts of the most severe kind, and to practice regular and daily, and sometimes thrice daily ablutions. They subsist much on green herbs, roots, and fruits; and at some periods of their ministry, they live much in the open air. And yet those of them who are true

Bramins—who live up to the dignity of their profession—are among the most healthy, vigorous, and long-lived of their race. The accounts of their longevity may, in some instances, be exaggerated; but it is certain that, other things being equal, they do not in this respect fall behind any other caste of their countrymen.

SOCIETY OF BIBLE CHRISTIANS.

This society has existed in Great Britain nearly half a century. They abstain from flesh, fish, and fowl—in short, from every thing that has animal life—and from all alcoholic liquors. Of their number in the kingdom I am not well informed. In Manchester they have three churches that have regular preachers; and frequent meetings have been held for discussing the diet question within a few years, some of which have been well attended, and all of which have been interesting. Among those who have adopted "the pledge" at their meetings, are some of the most distinguished men in the kingdom, and a few of the members of parliament. Through these and other instrumentalities, the question is fairly up in England, and will not cease to be discussed till fairly settled.

A branch or colony from the parent society, under the pastoral care of Rev. Wm. Metcalfe, consisting of only eight members, came in 1817 and established itself in Philadelphia. They were incorporated as a society in 1830. In 1846 the number of their church members was about seventy, besides thirty who adhered to their abstemious habits, but were not in full communion. During the thirty years ending in 1846, twelve of their number died—four children and eight adults. The average age of the latter was fifty-seven years. Of the seventy now belonging to the society, nineteen are between forty and eighty years of age; and forty, in all, over twenty-five. Of the whole number, twelve have abstained from animal food thirty-seven years, seven from twenty to thirty years, and fifty-one never tasted animal food or drank intoxicating drinks.

And yet they are all—if we except Mr. Metcalfe, their minister—of the laboring class, and hard laborers, too. Their strength and power of endurance is fully equal to their neighbors in similar circumstances, and in several instances considerably superior. Mr. Fowler, the phrenologist, testifies, concerning one of them, that he is regarded as the strongest man in Philadelphia. I have long had acquaintance with this sect, through Mr. M., of Philadelphia, and Mr. Simpson, one of their leading men in England, and have not a doubt of the truth of what has been publicly stated concerning them. They are a modest people, and make few pretensions; and yet they are a very meritorious people.

One thing very much to their advantage, as it shows the health-giving, health-preserving tendency of their practice and principles, remains to be related.

When the yellow fever prevailed in Philadelphia, in 1818 and 1819, the infection seemed specially rife in the immediate vicinity of the Bible Christians. So, also, in 1832, with the cholera. And yet none of them fled. There they remained during the whole period of suffering, and afforded their sick neighbors all the relief in their power. Their minister, in particular, was unwearied in his efforts to do good. Yet not one of their little number ever sickened or died of either yellow fever or cholera.

Till within a few years, they have been governed solely by regard to religious principle, having known little of Physiology or any other science bearing on health. Of late, however, they have turned their attention to the subject, and have among them a respectable Physiological society, which holds its regular meetings, and is said to be flourishing.

From one of their publications, entitled "Vegetable Cookery," I have extracted the following very brief summary of their views concerning the use of animals for sustenance.

"The Society of Bible Christians abstain from animal food, not only in obedience to the Divine command, but because it is an observance, which, if more generally adopted, would prevent much cruelty, luxury, and disease, besides many other evils which cause misery in society. It would be productive of much good, by promoting health, long life, and happiness, and thus be a most effectual means of reforming mankind. It would entirely abolish that greatest of curses, *war*; for those who are so conscientious as not to kill animals, will never murder human beings. On all these accounts the system cannot be too much recommended. The practice of abstaining cannot be wrong; it must therefore be some consolation to be on the side of duty. If we err, we err on the sure side; it is innocent; it is infinitely better authorized and more nearly associated with religion, virtue, and humanity, than the contrary practice—and we have the sanction of the wisest and the best of men—of the whole Christian world, for several hundred years after the commencement of the Christian era."

ORPHAN ASYLUM OF ALBANY.

I class this as a community, because it is properly so, and because I cannot conveniently class it otherwise. The facts which are to be related are too valuable to be lost. They were first published, I believe, in the Northampton Courier; and subsequently in the Boston Medical and Surgical Journal, and in the Moral Reformer. In the present case, the account is greatly abridged.

The Orphan Asylum of Albany was established about the close of the year 1829, or the beginning of the year 1830. Shortly after its establishment, it contained seventy children, and subsequently many more. The average number, from its commencement to August 1836, was eighty.

For the first three years, the diet of the inmates consisted of fine bread, rice, Indian puddings, potatoes, and other vegetables and fruits, with milk; to which was added flesh or flesh-soup once a day. Considerable attention was also paid to bathing and cleanliness, and to clothing, air, and exercise. Bathing, however, was performed in a perfect manner, only once in three weeks. As many of them were received in poor health, not a few continued sickly.

In the fall of 1833, the diet and regimen of the inmates were materially changed. Daily ablution of the whole body, in the use of the cold shower or sponge bath—or, in cases of special disease, the tepid bath was one of the first steps taken; then the fine bread was laid aside for that made of unbolted wheat meal; and soon after flesh and flesh-soups were wholly banished; and thus they continued to advance, till, in about three months more, they had come fully upon the vegetable system, and had adopted reformed habits in regard to sleeping, air, clothing, exercise, etc. On this course, then, they continued to August, 1836, and, for aught I know, to the present time. The results were as follows:

During the first three years, or while the old system was followed, from four to six children were continually on the sick list, and sometimes more; and one or two assistant nurses were necessary. A physician was needed once, twice, or three times a week, uniformly; and deaths were frequent. During this whole period there were between thirty and forty deaths.

After the new system was fairly adopted, the nursery was soon entirely vacated, and the services of the nurse and physician no longer needed; and for more than two years no case of sickness or death took place. In the succeeding twelve months there were three deaths, but they were new inmates, and were diseased when they were received; and two of them were idiots. The Report of the Managers says, "Under this system of dietetics (though the change ought not to be wholly attributed to the diet) the health of the children has not only been preserved, but those who came to the asylum weakly, have become healthy and strong, and greatly increased in activity, cheerfulness, and happiness." The superintendents also state, that "since the new regimen has been fully adopted, there has been a remarkable increase of health, strength, activity, vivacity, cheerfulness, and contentment among the children. Indeed, they appear to be, uniformly, perfectly healthy and happy; and the strength and activity they exhibit are truly surprising. The change of temper is very great. They have become less turbulent, irritable, peevish, and discontented; and far more manageable, gentle, peaceable, and kind to each other." One of them further observes, "There has been a great increase in their mental activity and power; the quickness and acumen of their perception, the vigor of their apprehension, and the power of their retention daily astonish me."

Such an account hardly needs comment; and I leave it to make its own impression on the candid and unbiassed mind and heart of the reader.

THE MEXICAN INDIANS.

The Indian tribes of Mexico, according to the traveler Humboldt, live on vegetable food. A spot of ground, which, if cultivated with wheat, as in Europe, would sustain only ten persons, and which by its produce, if converted into pork or beef, would little more than support one, will in Mexico, when used for banana, sustain equally well two hundred and fifty.

The reader will do well to take the above fact, and the estimates appended to it, along with him when he comes to examine what I have called the economical argument of the great diet question, in our last chapter, under the head, "The Moral Argument." We shall do well to remember another suggestion of Humboldt, that the habit of eating animals diminishes our natural horror of cannibalism.

SCHOOL IN GERMANY.

There is, in the Annals of Education for August, 1836, an account of a school in which the same simple system which was pursued in the Orphan Asylum at Albany was adopted, and with the same happy results. I say the *same* system; I believe plain meat was allowed occasionally, but it was seldom. Their food was exceedingly simple, consisting chiefly of bread and other vegetables, fruits and milk. Great attention was also paid to daily cold bathing. The following is the teacher's statement in regard to the results:

"I am at present the foster father of nearly seventy young people, who were born in all the varieties of climate from Lisbon to Moscow, and whose early education was necessarily very different. These young men are all healthy; not a single eruption is visible on their faces; and three years often pass, during which not a single one of them is confined to his bed; and in the twenty-one years that I have been engaged in this institution, not one pupil has died. Yet, I am no physician. During the first ten years of my residence here, no physician entered my house; and, not till the number of my pupils was very much increased, and I grew anxious not to overlook any thing in regard to them, did I begin to seek at all for medical advice.

"It is the mode of treating the young men here, which is the cause of their superior health; and this is the reason why death has not yet entered our doors. Should we ever deviate from our present principles—should we approach nearer the mode of living common in wealthy families—we should soon be obliged to establish, in our institution, as it is in others, medicine closets and nurseries. Instead of the freshness which now adorns the cheeks of our youth, paleness would appear, and our church-yards would contain the tombs of promising young men, who, in the bloom of their years, had fallen victims to disease."

THE AMERICAN PHYSIOLOGICAL SOCIETY.

This association was formed in 1837. When first formed, it consisted of one hundred and twenty-four males, and forty-one females; in all, one hundred and sixty-five. Their number soon increased to more than two hundred.

Most of these individuals were more or less feeble, and a very large proportion of them were actually suffering from chronic disease when they became members of the society. Not a few joined it, indeed, as a last resort, after having tried every thing else, as drowning men are said to catch at straws.

Nearly if not quite all the members of this society, as well as most of their families, abstained for a time from animal food. Some of them even adopted the vegetable system a year or so earlier. And there were a few who adopted it much sooner—one or two of them eight years earlier.

Of the individuals belonging to the Physiological Society or to their families, and adhering to the same principles, two adults only died, and one child, during the first two years. I will not be quite positive, but there were four in all, two adults, and two children; but this was the extent of mortality among them for about fifteen months.

The whole number of those who belonged to the society, with those members of their families who adhered to their principles (estimating families, as is usually done, at five members to each), is believed to have been from three hundred and twenty to three hundred and fifty. The average mortality for the same number of healthy persons, during the same period, in Boston and the adjacent places, was about six or seven; though in some places it was much greater. In a single parish in Roxbury—and without any remarkable sickness—the mortality, for the same number of persons, was equal to ten or twelve.

Now, we must not forget, what I have already stated, that this society of vegetable-eaters—the two hundred adults, I mean—were generally invalids, and some of them given over by physicians. Instead, therefore, of only half the usual proportion of deaths among them, we might naturally enough have expected twice or three times the usual number. And this expectation would have appeared still better founded when it was considered that many made the change in their habits, and especially in their diet, very suddenly.

But the whole story is not yet told. Not only was the number of deaths very small, as above stated, but there were a great number of remarkable recoveries. Some, who had very obstinate complaints, appeared, for a time, to be entirely well. Others were getting well as fast as could be expected. Some, who were broken down and prematurely old, seemed to renew their youth. Many became free from colds and eruptive complaints, to which they were formerly subject. And those who had acute diseases, of whom, however, the number was very small, did not suffer so much as is usually the case with flesh-eaters in circumstances otherwise apparently similar.

But a reverse at length came. They were led into their abstemious course by mere impulse in very many cases, and though a library was formed and meetings held, nobody, hardly, would read, and the meetings grew thin. They had no Joe Smith or Gen. Taylor to lead them—and mankind without leaders and without deep-toned principle, soon grow tired of war. Few will fight in such circumstances.

CHAPTER VIII.

VEGETABLE DIET DEFENDED.

General Remarks on the Nature of the Argument—1. The Anatomical Argument.—2. The Physiological Argument.—3. The Medical Argument.—4. The Political Argument.—5. The Economical Argument.—6. The Argument from Experience.—7. The Moral Argument.—Conclusion.

In the progress of a work like this, it may not be amiss to present, in a very brief manner, the general arguments in defence of a diet exclusively vegetable. Some of them have, indeed, already been adverted to in the testimony of the preceding chapters; but not all. Besides, it seemed to me desirable to collect the whole in a general view.

There are various ways of doing this, according to the different aspects in which the subject is viewed. Every one has his own point of observation. I have mine. Conformably to the view I have taken, therefore, I shall endeavor to arrange my remarks under the nine following heads, viz., the ANATOMICAL, the PHYSIOLOGICAL, the MEDICAL, the POLITICAL, the ECONOMICAL, the EXPERIMENTAL, the MORAL, the MILLENNIAL, and the BIBLE ARGUMENTS.

Dr. Cheyne relied principally on what I have called the medical argument—though what I mean by this may not be quite obvious, till I shall have presented it in its proper place. Not that he wholly overlooked any thing else; but this, as it seems to me, was with him the grand point. Nearly the same might be said of Dr. Lambe, and of several others.

Dr. Mussey seems to place the anatomical and physiological arguments in the foreground. It is true he makes much use of the medical and the moral arguments; but the former appear to be his favorites. Dr. Whitlaw, and some others, incline to make the moral and political arguments more prominent. Mr. Graham, who has probably done more to reduce the subject of vegetable dietetics to a *system* than any other individual,—though he makes much use of *all* the rest, especially the moral and medical,—appears to dwell with most interest on the physiological argument. This seems to be, with him, the strong-hold—the grand citadel. And it must be confessed that the point of defence is very strong indeed, as we shall see in the sequel.

If I have a favorite, with the rest, it is the moral argument, or perhaps a combination of this with the economical. But then I dwell on the latter with so much interest, chiefly on account of the former. I would give very little to be able to bring the world of mankind back to nature's true simplicity, if it were only to make them better and more perfect animals; though I know not but an attempt of this sort would be as truly laudable as the attempt so often made to improve the breed of our domestic animals. I suppose man, considered as a mere animal, is superior, in point of importance to all the others. But, after all, I would reform his dietetic habits principally to make him better, morally; to make him better, in the discharge of his varied duties to his fellow-beings and to God. I would elevate him, that he may become as truly god-like, or godly as he now too often is, by his unnatural habits, earthly or beastly. I would render him a rational being, fitted to fill the space which he appears to have been originally designed to fill—the gap in the great chain of being between the higher quadrupeds and the beings we are accustomed to regard as angelic. I would restore him to his true dignity. I would make him a child of God, and an *heir* of a glorious immortality.

But I now proceed to the discussion of the subject which I have assigned to this chapter.

I. THE ANATOMICAL ARGUMENT.

There has been a time when the teeth and intestines of man were supposed to indicate the necessity of a mixed diet—a diet partly animal and partly vegetable. Four out of thirty-two teeth were found to resemble slightly, the teeth of carnivorous animals. In like manner, the length of the intestinal tube was thought to be midway between that of the flesh-eating, and that of the herb-eating quadrupeds. But, unfortunately for this mode of defending an animal diet, it has been found out that the fruit and vegetable-eating monkey race, and the herb-eating camel, have the said four-pointed teeth much more pointed than those of man and that the intestines, compared with the real length of the body, instead of assigning to man a middle position, would place him among the herbivorous animals. In short—for I certainly need not dwell on this part of my subject, after having adduced so fully the views of Prof. Lawrence and Baron Cuvier—there is no intelligent naturalist or comparative anatomist, at present, who attempts to resort for one moment to man's structure, in support of the hypothesis that he is a flesh-eater. None, so far as I know, will affirm, or at least with any show of reason maintain, that anatomy, so far as that goes, is in favor of flesh eating. We come, then, to another and more important division of our subject.

II. THE PHYSIOLOGICAL ARGUMENT.

One of the advantages of vegetable-eaters over others, is in the superior appetite which they enjoy. There are many flesh-eaters who have what they call a good appetite. But I never knew a person of this description, who made the change from a mixed diet to one purely vegetable, who did not afterward acknowledge that he never once knew, while he was an eater of animal food, a truly perfect appetite. This testimony in favor of vegetable diet is positive; whereas that of the multitude, who have never made the change I speak of, but who are therefore the more ready to laugh at the conclusions, is merely negative.

A person of perfect appetite can eat at all times, and under all circumstances. He can eat of one thing or another, and in greater or less quantity. Were there no objections to it, he could make an entire meal of the coarsest and most indigestible substances; or, he could eat ten or fifteen times a day; or, he could eat a quantity at once which would astonish even a Siberian; or, on the contrary, he could abstain from food entirely, for a short time; and any of these without serious inconvenience. He would, indeed, feel a slight want of something (in the case of total abstinence), when the usual hour arrived for taking a meal; but the sensation is not an abiding one; when the hour has passed by, it entirely disappears. Nor is there ever, at least for a day or two of abstinence, that gnawing at the stomach, as some express it, which is so often felt by the flesh-eater and the devourer of other mixed and injurious dishes and which is so generally mistaken for true and genuine hunger.

I have said that the vegetable-eater finds no serious inconvenience from the quality or quantity of his food; but I mean to speak here of the *immediate* effects solely. No doubt every error of this sort produces mischief, sooner or later. The more perfect the appetite is, the greater should be our moral power of commanding it, and of controlling the quality and quantity of our food and drink, as well as the times and seasons of receiving it.

These statements, I am aware, are contrary to the received and current opinion; but that they are true, can be proved, not by one person merely,— though if that person were to be entirely relied on, his positive affirmation would outweigh a thousand *negative* testimonies,—but by many hundreds. It is more generally supposed that he who confines himself to a simple diet, soon brings his stomach into such a state that the slightest departure from his usual habits for once only, produces serious inconveniences; and this indeed is urged as an argument against simplicity itself. Yet, how strange! How much more natural to suppose that the more perfect the health of the stomach, the better it will bear, for a time, with slight or even serious departures from truth and nature! How much more natural to suppose that perfect health is the very best defence against all the causes which tend to invite or to provoke disease! And what it would be natural to infer, is proved by experience to be strictly true. The thorough-going vegetable-eater can make a meal for once, or perhaps feed for a day or so, on substances which would almost kill many others; and can do so with comparative impunity. He can make a whole meal of cheese, cabbage, fried pudding, fried dough-nuts, etc., etc.; and if it be not in remarkable excess, he will feel no immediate inconvenience, unless from the mental conviction that he must pay the full penalty at some distant day.

I repeat it, the appetite of the vegetable-eater, if true to his principles, and temperate in regard to quantity, is always, at all moments of his life, perfect. To be sure, he is not always *hungry*. Hunger, indeed, as I have already intimated—what most people call hunger, a morbid sensation, or gnawing—is unknown to him. But there is scarce a moment of his life, at least, when he is awake, in which he could not enjoy the pleasures of eating, even the coarsest viands, with a high relish; provided, however, he knew it was *proper* for him to eat. Nor is his appetite fickle, demanding this or that particular article, and disconcerted if it cannot be obtained. It is satisfied with any thing to which the judgment directs; and though gratified, in a high degree, with dainties, when nothing better and more wholesome cannot be obtained, never demanding them in a peremptory manner.

The vegetable-eater has a more quiet, happy, and perfect digestion than the flesh-eater. On this point there has been much mistake, even among physiologists. Richerand and many others suppose that a degree of constitutional disturbance is indispensable during the process of digestion; and some have even said that the system was subjected at every meal—nay, at every healthy meal—to a species of miniature fever. The remarks of Richerand are as follows. I have slightly abridged them, but have not altered the sense:

"While the alimentary solution is going on, a slight shivering is felt; the pulse becomes quicker and more contracted; the vital power seems to forsake the other organs, to concentrate itself on that which is the seat of the digestive process. As the stomach empties itself, the shivering is followed by a gentle warmth; the pulse increases in fullness and frequency; and the insensible perspiration is augmented. Digestion brings on, therefore, a general action, analogous to a febrile paroxysm."

And what is it, indeed, *but* a febrile paroxysm? Nay, Richerand himself confirms this by adding, "this fever of digestion, noticed already by the ancients, is particularly observable in women of great sensibility." That is, the fever is more violent in proportion to the want of power in the person it attacks to resist its influence; just as it is with fever in all other circumstances, or when induced by any other causes.

But, can any one believe the Author of Nature has so made us, that in a steady and rational obedience to his laws, it is indispensable that we should be thrown into a fever three times a day, one thousand and ninety-five times in a year, and seventy-six thousand six hundred and fifty in seventy years? No wonder, if this were true, that the vitality of our organs was ordained to wear out soon; for we see by what means the result would be accomplished.

The fever, however, of which Richerand speaks, does very generally exist, because mankind very generally depart from nature and her laws. But it is not necessary. The simple vegetable-eater—if he lives right in all other respects—if he errs not as to quantity, knows nothing of it; nor should it be known by any body. We should leave it to the animals below man to err, in quantity and quality, to an excess which constitutes a surfeit or a fever, and causes fullness and drowsiness, and a recumbent posture. The self-styled lord of the animal world should rise superior to habits which have marked, in every age, certain orders of the lower animals.

But the chyle which is produced from vegetable aliment is better—all other things being equal—than that which is produced from any other food. For proof of this, we need but the testimony of Oliver and other physiologists. They tell us, unhesitatingly, that under the same circumstances, chyle which is formed from vegetables will be preserved from putrefaction many days longer—the consequence of greater purity and a more perfect vitality—than that which is formed from any admixture of animal food. Is it not, then, better for the purposes of health and longevity? Can it, indeed, be otherwise? I will say nothing at present, for want of space to devote to it, of the indications which are afforded by the other sensible properties of the chyle which is produced from vegetables. The single fact I have presented is enough on that point.

The best solids and fluids are produced by vegetable eating. On this single topic a volume might be written, without exhausting it, while I must confine myself to a page or two.

In the first place, it forms better bones and more solid muscles, and consequently gives to the frame greater solidity and strength. Compare, in evidence of the truth of this statement, the vegetable-eating millions of middle and southern Europe, with the other millions, who, supposed to be more fortunate, can get a little flesh or fish once a day. Especially, make this comparison in Ireland, where the vegetable food selected is far from being of the first or best order; and whose sight is so obtuse as not to perceive the difference? I do not say, compare the enervated inhabitant of a hot climate, as Spain or Italy, with the inhabitant of England, or Scotland, or Russia, for that would be an unfair comparison, wholly so; but compare Italian with Italian, Frenchman with Frenchman, German with German, Scotchman with Scotchman, and Hibernian with Hibernian.

In like manner, compare the millions of Japanese of the interior, who subsist through life chiefly on rice, with the few millions of the coasts who eat a little fish with their rice. Make a similar comparison in China and in Hindostan. Notice, in particular, the puny Chinese, who live in southern China, on quite a large proportion of shell-fish, compared with the Chinese of the interior. Extend your observations to Hindostan. Do not talk of the effeminate habits and weak constitutions of the rice and curry eaters there—bad as the admixture of rice and curry may be—for that is to compare the Hindoo with other nations; but compare Hindoo with Hindoo, which is the only fair way. Compare the porters of the Mediterranean, both of Asia and Europe, who feed on bread and figs, and carry weights to the extent of eight hundred or one thousand pounds, with the porters who eat flesh, fish, and oil. Compare African with African, American Indian with American Indian; nay, even New Englander with New Englander; for we have a few here who are trained to vegetable eating. In short, go where you will, and institute a fair comparison, and the results will be, without a single exception, in favor of a diet exclusively vegetable. It is necessary, however, in making the comparison, to place *good* vegetable food in opposition to good animal food; for no one will pretend that a diet of crude, miserable, or imperfect, or sickly vegetables will be as wholesome as one consisting of rich farinaceous articles and fruits; nor even as many kinds of plain meat.

The only instance which, on a proper comparison, will probably be adduced to prove the incorrectness of these views, will be that of a few tribes of American Indians, who, though they have extremely robust bodies, are eaters of much flesh. But they live also in the open air, and have many other good habits, and are healthy in spite of the inferiority of their diet. But perfect, physically, as they seem to be, and probably are, examine the vegetable-eaters among them, of the same tribe, and they will be found still more so.

In the next place, the fluids are all in a better and more healthy state. In proof of this, I might mention in the first place that superior agility, ease of motion, speed, and power of endurance which so distinguish vegetable-eaters, wherever a fair comparison is instituted. They possess a suppleness like that of youth, even long after what is called the juvenile period of life is passed over. They are often seen running and jumping, unless restrained by the arbitrary customs of society, in very advanced age. Their wounds heal with astonishing rapidity in as many days as weeks, or even months, in the latter case. All this could not happen, were there not a good state of the fluids of the system conjoined, to a happy state of the solids.

The vegetable-eater, if temperate in the use of his vegetables, and if all his other habits are good, will endure, better than the flesh-eater, the extremes of heat and cold. This power of endurance has ever been allowed to be a sure sign of a good state of health. The most vigorous man, as it is well known, will endure best both extremes of temperature. But it is a proof also of the greater purity of his solids and fluids.

The secretions and excretions of his body are in a better state; and this, again, proves that his blood and other fluids are healthy. He does not so readily perspire excessively as other men, neither is there any want of free and easy perspiration. Profuse sweating on every trifling exertion of the body or mind, is as much a disease as an habitually dry skin. But the vegetable-eater escapes both of these extremes. The saliva, the tears, the milk, the gastric juice, the bile, and the other secretions and excretions—particularly the dejections—are as they should be. Nay, the very exhalations of the lungs are purer, as is obvious from the breath. That of a vegetable-eater is perfectly sweet, while that of a flesh-eater is often as offensive as the smell of a charnel-house. This distinction is discernible even among the brute animals. Those which feed on grass, grain, etc., have a breath incomparably sweeter than those which prey on animals. Compare the camel, and horse, and cow, and sheep, and rabbit, with the tiger (if you choose to approach him), the wolf, the dog, the cat, and the hawk. One comparison will be sufficient; you will never forget it. But there is as much difference between the odor of the breath of a flesh-eating human being and a vegetable-eater, as between those of the dog and the lamb. This, however, is a secret to all but vegetable-eaters themselves, since none but they are so situated as to be able to make the comparison. But, betake yourself to mealy vegetables and fruits a few years, and live temperately on them, and then you will perceive the difference, especially in riding in a stage-coach. This, I confess, is rather a draw-back upon the felicity of vegetable-eaters; but it is some consolation to know what a mass of corruption we ourselves have escaped.

There is one more secretion to which I wish to direct your attention, which is, the fat or oil. The man who lives rightly, and rejects animal food among the rest, will never be overburdened with fat. He will neither be too corpulent nor too lean. Both these conditions are conditions of disease, though, as a general rule, corpulence is most to be dreaded; it is, at least, the most disgusting. Fat, I repeat it, is a secretion. The cells in which it is deposited serve for relieving the system of many of the crudities and abuses, not to say poisons, which are poured into it—cheated; as it were, in some degree into the blood, secreted into the fat cells, and buried in the fat to be out of the way, and where they can do but little mischief. Yet, even here they are not wholly harmless. The fat man is almost always more exposed to disease, and to *severe* epidemic disease in particular, than the lean man. Let us leave it to the swine and other kindred quadrupeds, to dispose of gross half poisonous matter, by converting it into, or burying it in fat; let us employ our vital forces and energies in something better. Above all, let us not descend to swallow, as many have been inclined to do, besides the ancient Israelites, this gross secretion, and reduce ourselves to the painful necessity of carrying about, from day to day, a huge mass of double-refined disease, pillaged from the foulest and filthiest of animals.

Vegetable-eaters—especially if they avoid condiments, as well as flesh and fish—are not apt to be thirsty. It is a common opinion among the laboring portion of the community, that they who perspire freely, must drink freely.

And yet I have known one or two hard laborers who were accustomed to sweat profusely and freely, who hardly ever drank any thing, except a little tea or milk at their meals, and yet were remarkably strong and healthy, and attained to a great age. One of this description (Frederick Lord, of Hartford, Conn.), lived to about the age of eighty-five. How the system is supplied, in such cases, with fluid, I do not know; but I know it is not necessary to drink perpetually for the purpose; for if but one healthy man can dispense with drinking, others may. The truth is, we seldom drink from real thirst. We drink chiefly either from habit, or because we have created a morbid or diseased thirst by improper food or drink, among which animal food is pretty conspicuous.

I have intimated that, in order to escape thirst, the vegetable-eater must abstain also from condiments. This he will be apt to do. It is he who eats flesh and fish, and drinks something besides water, who feels such an imperious necessity for condiments. The vegetable and milk eater, and water-drinker, do not need them.

It is in this view, that the vegetable system lies at the foundation of all reform in the matter of temperance. So long as the use of animal food is undisturbed and its lawfulness unquestioned, all our efforts to heal the maladies of society are superficial. The wound is not yet probed to the bottom. But, renounce animal food, restore us to our proper condition, and feed us on milk and farinaceous articles, and our fondness for excitement and our hankering for exciting drinks and condiments will, in a few generations, die away. Animal food is a root of all evil, so far as temperance is concerned, in its most popular and restricted sense.

The pure vegetable-eaters, especially those who are trained as such, seldom drink at all. Some use a little water with their meals, and a few drink occasionally between them, especially if they labor much in the open air, and perspire freely. Some taste nothing in the form of drink for months, unless we call the abundant juices of apples and other fruits, and milk, etc., by that name—of which, by the way, they are exceedingly fond. The reason is, they are seldom thirsty. Dr. Lambe, of London, doubts whether man is naturally a drinking animal; but I do not carry the matter so far. Still I believe that ninety-nine hundredths of the drink which is used, *as* now used, does more harm than good.

He who avoids flesh and fish, escapes much of that languor and faintness, at particular hours, which others feel. He has usually a clear and quiet head in the morning. He is ready, and willing, and glad to rise in due season; and his morning feelings are apt to last all day. He has none of that faintness between his meals which many have, and which tempts thousands to luncheons, drams, tobacco, snuff, and opium, and ultimately destroys so much health and life. The truth is, that vegetable food is not only more quiet and unstimulating than any other, but it holds out longer also. I know the contrary of this is the general belief; but it is not well founded. Animal food stimulates most, and as the stimulus goes off soon, we are liable to feel dull after it, and to fancy we need the stimulus of drink or something else to keep us up till the arrival of another meal. And, having acquired a habit of relying on our food to stimulate us immediately, much more than to give us real, lasting, permanent strength, it is no wonder we feel, for a time, a faintness if we discontinue its use. This only shows the power of habit, and the over-stimulating character of our accustomed food. Nor does the simple vegetable-eater suffer, during the spring, as other people say they do. All is cheerful and happy with him, even then. Nor, lastly, is he subject to hypochondria or depression of spirits. He is always lively and cheerful; and all with him is bright and happy. As it has been expressed elsewhere, with the truly temperate man it is "morning all day."

The system of diet in question, greatly improves, exalts, and perfects the senses. The sight, smell, and taste are rendered greatly superior by it. The difference in favor of the hearing and the touch may not be so obvious; nevertheless, it is believed to be considerable. But the change in the other senses—the first three which I have named—even when we reform as late as at thirty-five or forty, is wonderful. I do not wish to encourage, by this, a delay of the work of reformation; we can never begin it too early.

Vegetable diet favors beauty of form and feature. The forms of the natives of some of the South Sea Islands, to say nothing of their features, are exceedingly fine. They are tall and well proportioned. So it is with the Japanese and Chinese, especially of the interior, where they subsist almost wholly on rice and fruits. The Japanese are the finest men, physically speaking, in Asia. The New Hollanders, on the contrary, who live almost wholly on flesh and fish, are among the most meagre and ugly of the human race, if we except the flesh-eating savages of the north, and the Greenlanders and Laplanders. In short, the principle I have here advanced will hold, as a *general rule*, I believe, other things being equal, throughout the world. If it be asked whether I would exalt beauty and symmetry into virtues, I will only say that they are not without their use in a virtuous people; and I look forward to a period in the world's history, when all will be comparatively well formed and beautiful. Beauty is exceedingly influential, as every one must have observed who has been long in the world; at least, if he has had his eyes open. And it is probably right that it should be so. Our beauty is almost as much within our control, as a race, as our conduct.

A vegetable diet, moreover, promotes and preserves a clearness and a generally healthful state of the mental faculties. I believe that much of the moral as well as intellectual error in the world, arises from a state of mind which is produced by the introduction of improper liquids and solids into the stomach, or, at least, by their application to the nervous system. Be this as it may, however, there is nothing better for the brain than a temperate diet of well-selected vegetables, with water for drink. This Sir Isaac Newton and hundreds of others could abundantly attest.

It also favors an evenness and tranquillity of temper, which is of almost infinite value. The most fiery and vindictive have been enabled, by this means, when all other means had failed, to transform themselves into rational beings, and to become, in this very respect, patterns to those around them. If this were its only advantage, in a physiological point of view, it would be of more value than worlds. It favors, too, simplicity of character. It makes us, in the language of the Bible, to remain, or to become, as little children, and it preserves our juvenile character and habits through life, and gives us a green old age.

Finally and lastly, it gives us an independence of external things and circumstances, that can never be attained without it. In vain may we resort to early discipline and correct education—in vain to moral and religious training—in vain, I had almost said, to the promises and threatenings of heaven itself, so long as we continue the use of food so unnatural to man as the flesh of animals, with the condiments and sauces, and improper drinks which follow in its train. Our hope, under God, is, in no small degree, on a radical change in man's dietetic habits—in a return to that simple path of truth and nature, from which, in most civilized countries, those who have the pecuniary means of doing it have unwisely departed.

III. THE MEDICAL ARGUMENT.

If perfect health is the best preventive and security against disease, and if a well-selected and properly administered vegetable diet is best calculated to promote and preserve that perfect health, then this part of the subject—what I have ventured to call the medical argument—is at once disposed of. The superiority of the diet I recommend is established beyond the possibility of debate. Now that this is the case—namely, that this diet is best calculated to promote perfect health—I have no doubt. For the sake of others, however, it may be well to adduce a few facts, and present a few brief considerations.

It is now pretty generally known, that Howard, the philanthropist, was, for about forty years a vegetable-eater, subsisting for much of this time on bread and tea, and that he went through every form of exposure to disease, contagious and non-contagious, perfectly unharmed. And had it not been for other physical errors than those which pertain to diet, I know of no reason why his life might not have been preserved many years longer—perhaps to this time.

Rev. Josiah Brewer, late a missionary in Smyrna, was very much exposed to disease, and, like Mr. Howard, to the plague itself; and yet I am not aware that he ever had a single sick day as the consequence of his exposure. I do not know with certainty that he abstains entirely from flesh meat, but he is said to be rigidly temperate in other respects.

Those who have read Rush's Inquiries and other writings, are aware that he was very much exposed to the yellow fever in Philadelphia, during the years in which it prevailed there. Now, there is great reason for believing that he owed his exemption from the disease, in part, at least, to his great temperance.

Mr. James, a teacher in Liberia, in Africa, had abstained for a few years from animal food, prior to his going out to Africa. Immediately after his arrival there, and during the sickly season, one of his companions who went out with him, died of the fever. Mr. James was attacked slightly, but recovered.

Another vegetable-eater—the Rev. Mr. Crocker—went out to a sickly part of Africa some years since, and remained at his station a long time in perfect health, while many of his friends sickened or died. At length, however, he fell.

Gen. Thomas Sheldon, of this state, a vegetable-eater, spent several years in the most sickly parts of the Southern United States, with an entire immunity from disease; and he gives it as his opinion that it is no matter where we are, so that our dietetic and other habits are correct.

Mr. G. McElroy, of Kentucky, spent several months of the most sickly season in the most unhealthy parts of Africa, in the year 1835, and yet enjoyed the best of health the whole time. While there and on his passage home, he abstained wholly from animal food, living on rice and other farinaceous vegetables and fruits.

In view of these facts and many others, Mr. Graham remarks: "Under a proper regimen our enterprising young men of New England may go to New Orleans or Liberia, or any where else they choose, and stay as long as they choose, and yet enjoy good health." And there is no doubt he is right.

But it is hardly worth while to cite single facts in proof of a point of this kind. There is abundant testimony to be had, going to show that a vegetable diet is a security against disease, especially against epidemics, whether in the form of a mere influenza or malignant fever. Nay, there is reason to believe that a person living according to *all* the Creator's laws, physical and moral, could hardly receive or communicate disease of any kind. How could a person in perfect health, and obeying to an iota all the laws of health—how could he contract disease? What would there be in his system which could furnish a nidus for its reception?

I am well aware that not a few people suppose the most healthy are as much exposed to disease as others, and that there are some who even suppose they are much more so. "Death delights in a shining mark," or something to this effect, is a maxim which has probably had its origin in the error to which I have adverted. To the same source may be traced the strange opinion that a fatal or malignant disease makes its first and most desperate attacks upon the healthy and the robust. The fact is—and this explains the whole riddle—those who are regarded, by the superficial and short-sighted in this matter, as the most healthy and robust, are usually persons whose unhealthy habits have already sown the seeds of disease; and nothing is wanting but the usual circumstances of epidemics to rouse them into action. More than all this, these strong-looking but inwardly-diseased persons are almost sure to die whenever disease does attack them, simply on account of the previous abuses of their constitutions.

During the prevalence of the cholera in New York, about the year 1832, all the Grahamites, as they were called, who had for some time abstained from animal food—and their number was quite respectable—and who persevered in it, either wholly escaped the disease, or had it very lightly; and this, too, notwithstanding a large proportion of them were very much exposed to its attacks, living in the parts of the city where it most prevailed, or in families where others were dying almost daily. This could not be the result of mere accident; it is morally impossible.

But flesh-eaters—admitting the flesh were wholesome—are not only much more liable to contract disease, but if they contract it, to suffer more severely than others. There is yet another important consideration which belongs to the medical argument. Animal food is much more liable than vegetable food, to those changes or conditions which we call poisonous, and which are always, in a greater or less degree, the sources of disease; it is also more liable to poisonous mixtures or adulterations.

It is true, that in the present state of the arts, and of agriculture and civic life generally, vegetables themselves are sometimes the sources of disease. I refer not to the spurred rye and other substances, which occasionally find their way into our fields and get mixed with our grains, etc., and which are known to be very active poisons,—so much as to the acrid or otherwise improper juices which are formed by forced vegetation, especially about cities, whether by means of hot-beds, green-houses, or new, strong, or highly-concentrated manures. I refer also to the crude, unripe, and imperfect fruits and other things with which our markets are filed now-a-days; and especially to *decaying* fruits and vegetables. But I cannot enlarge; a volume would be too little to do this part of the subject justice. Nothing is more wanted than light on this subject, and a consequent reform in our fashionable agriculture and horticulture.

And yet, although I admit, most cheerfully, the danger we are in of contracting disease by using diseased vegetables, the danger is neither so frequent nor so imminent, in proportion to the quantity of it consumed, as from animal food. Let us briefly take a view of the facts.

Milk, in its nature, approaches nearest to the line of the vegetable kingdom, and is therefore, in my view, the least objectionable form of animal food. I am even ready to admit that for persons affected with certain forms of chronic disease, and for all children, milk is excellent. And yet, excellent as it is, it is very liable to be injurious. We are told, by the most respectable medical men of France, that all the cows about Paris have tubercles (the seeds or beginning of consumption) in their lungs which is probably owing to the unnatural state in which they are kept, as regards the kind, and quantity, and hours of receiving their food; and especially as regards air, exercise, and water. Cows cannot be healthy, nor any other domestic animals, any more than men, when long subjected to the unnatural and unhealthy influences of bad air, want of exercise, etc. Hence, then, most of our cows about our towns and cities must be diseased, in a greater or less degree—if not with consumption, with something else. And of course their milk must be diseased—not, perhaps, as much as their blood and flesh, but more or less so. But if milk is diseased, the butter and cheese made from it must be diseased also.

But milk is sometimes diseased through the vegetables which are eaten by the cow. Every one knows how readily the sensible properties of certain acrid plants are perceived in the milk. Hence as I have elsewhere intimated, we are doubly exposed to danger from eating animal food; first, from the diseases of the animal itself, and secondly, from the diseases which are liable to be induced upon us by the vegetables they use, some of which are not poisonous to them, but are so to us. So that, in avoiding animal food, we escape at least a part of the danger.

Besides the general fact, that almost all medical and dietetic writers object to fat, and to butter among the rest, as difficult of digestion and tending to cutaneous and other diseases,—and besides the general admission in society at large that it makes the skin "break out,"—it must be obvious that it is liable to retain, in a greater or less degree, all the poisonous properties which existed in the milk from which it was made. Next to fat pork, butter seems to me one of the worst things that ever entered a human stomach; and if it will not, like pork, quite cause the leprosy, it will cause almost every other skin disease which is known.

Cheese is often poisoned now-a-days by design. I do not mean to say that the act of poisoning is accompanied by malice toward mankind; far from it.

It is added to color it, as in the form of anatto; or to give it freshness and tenderness, as in the case of arsenic.[21]

Eggs, when not fresh, are more or less liable to disease. I might even say more. When not fresh, they *are* diseased. On this point we have the testimony of Drs. Willich and Dunglison. The truth is, that the yolk of the egg has a strong tendency to decomposition, and this decomposing or putrefying process *begins* long before it is perceived, or even suspected, by most people. There is much reason for believing that a large proportion of the eggs eaten in civic life,—except when we keep the poultry ourselves,—are, when used, more or less in a state of decomposition. And yet, into how many hundred forms of food do they enter in fashionable life, or in truth, in almost every condition of society! The French cooks are said to have six hundred and eighty-five methods of cooking the egg, including all the various sorts of pastry, etc., of which it forms a component part.

One of the grand objections against animal food, of almost all sorts, is, that it tends with such comparative rapidity to decomposition. Such is at least the case with eggs, flesh, and fish of every kind. The usual way of preventing the decomposition is by processes scarcely less hurtful—by the addition of salt, pyroligneous acid, saltpetre, lime, etc. These, to be sure, prevent putrefaction; but they render every thing to which they are applied, unless it is the egg, the more indigestible.

It is a strange taste in mankind, by the way, which leads them to prefer things in a state of incipient decomposition. And yet such a taste certainly prevails widely. Many like the flesh beaten; hence the origin of the cruel practice of the East of whipping animals to death.[22] And most persons like fresh meat kept till it begins to be *tender*; that is, begins to putrefy. So most persons like fermented beer better than that which is unfermented, although fermentation is a step toward putrefaction; and they like vinegar, too, which is also far advanced in the same road.

That diseased food causes diseases in the persons who use it, needs not, one would think, a single testimony; and yet, I will name a few.

Dr. Paris, speaking of fish, says,—"It is not improbable that certain cutaneous diseases may be produced, or at least aggravated by such diet." Dr. Dunglison says, bacon and cured meats are often poisonous. He speaks of the poisonous tendency of eggs, and says that all *made* dishes are more or less "rebellious." In Aurillac, in France, not many years since, fifteen or sixteen persons were attacked with symptoms of cholera after eating the milk of a certain goat. The goat died with cholera about twenty-four hours after, and two men, no less eminent than Professors Orfila and Marc, gave it as their undoubted opinion that the cholera symptoms alluded to, were caused by the milk. I have myself known oysters at certain times and seasons to produce the same symptoms. During the progress of a mortal disease among the poultry on Edisto Island, S. C., in 1837, all the dogs and vultures that tasted of the flesh of the dead poultry sickened and died. Chrisiston mentions an instance in which five persons were poisoned by eating beef; and Dunglison one in which fourteen persons were made sick, and some died, from eating the meat of a calf. Between the years 1793 and 1827, it is on record that there were in the kingdom of Wurtemberg alone, no less than two hundred and thirty-four cases of poisoning, and one hundred and ten deaths, from eating sausages. But I need not multiply this sort of evidence, the world abounds with it; though for one person who is poisoned so much as to be made sick immediately, hundreds perhaps are only slightly affected; and the punishment may seem to be deferred for many years.

The truth, in short, is, that every fashionable process of fattening and even of domesticating animals, induces disease; and as most of the animals we use for food are domesticated or fattened, or both, it follows that most of our animal food, whether milk, butter, cheese, eggs, or flesh, is diseased food, and must inevitably, sooner or later, induce disease in those who receive it. Those which are most fattened are the worst, of course; as the hog, the goose, the sheep, and the ox. The more the animal is removed from a natural state, in fattening, the more does the fat accumulate, and the more it is diseased. Hence the complaints against every form of animal oil or fat, in every age, by men who, notwithstanding their complaints, for the most part, continue to set mankind an example of its use.

Let me here introduce a single paragraph from Dr. Cheyne, which is very much to my present purpose.

"About London, we can scarce have any but crammed poultry or stall-fed butchers' meat. It were sufficient to disgust the stoutest stomach to see the foul, gross, and nasty manner in which, and the fetid, putrid, and unwholesome materials *with* which they are fed. Perpetual foulness and cramming, gross food and nastiness, we know, will putrefy the juices, and corrupt the muscular substance of human creatures—and sure they can do no less in brute animals—and thus make our food poison. The same may be said of hot-beds, and forcing plants and vegetables. The only way of having sound and healthful animals, is to leave them to their own natural liberty in the free air, and their own proper element, with plenty of food and due cleanliness; and a shelter from the injuries of the weather, whenever they have a mind to retire to it."

The argument then is, that, for healthy adults at least, a well-selected vegetable diet, other things being equal, is a preventive of disease, and a security against its violence, should it attack us, in a far greater degree than a diet which includes animal food in any of its numerous forms. It will either prevent the common diseases of childhood, including those which are deemed contagious, or render their attacks extremely mild: it will either prevent or mitigate the symptoms of the severe diseases of adults, not excepting malignant fevers, small-pox, plague, etc.; and it will either prevent such diseases as cancer, gout, epilepsy, scrofula, and consumption, or prolong life under them.

Who that has ever thought of the condition of our domestic animals, especially about towns and cities—their want of good air, abundant exercise, good water, and natural food, to say nothing of the butter-cup and the other poisonous products of over-stimulating or fresh manures which they sometimes eat—has not been astonished to find so little disease among us as there actually is? Animal food, in its best state, is a great deal more stimulating and heating to the system than vegetable food;—but how much more injurious is it made, in the circumstances in which most animals are placed? Do we believe that even a New Zealand cannibal would willingly eat flesh, if he knew it was from an animal that when killed was laboring under a load of liver complaint, gout, consumption, or fever? And yet, such is the condition of most of the animals we slay for food. They would often die of their diseases if we did not put the knife to their throats to prevent it.

One more consideration. If the exclusive use of vegetable food will prevent a multitude of the worst and most incurable diseases to which human nature, in other circumstances, seems liable; if it will modify the diseases which a mixed diet, or absolute intemperance, or gluttony had induced,—by what rule can we limit its influence? How know we that what is so efficacious in regard to the larger diseases, will not be equally so in the case of all smaller ones? And why, then, may not its universal adoption, after a few generations, banish disease entirely from the world? Every person of common observation, knows that, as a general rule, they who approach the nearest to a pure vegetable and water diet, are most exempt from disease, and the longest-lived and most happy. How, then, can it otherwise happen than that a still closer approximation will afford a greater exemption still, and so on indefinitely? At what point of an approach toward such diet and regimen, and toward perfect health at the same time, is it that we stop, and more temperance still will injure us? In short, where do we cross the line?

IV. THE POLITICAL ARGUMENT.

I have dwelt at such length on the physiological and medical arguments in defence of the vegetable system, that I must compress my remaining views into the smallest space possible; especially those which relate to its political, national, or general advantages.

Political economists tell us that the produce of an acre of land in wheat, corn, potatoes, and other vegetables, and in fruits, will sustain animal life sixteen times as long as when the produce of the same acre is converted into flesh, by feeding and fattening animals upon it.

But, if we admit that this estimate is too high, and if the real difference is only eight to one, instead of sixteen to one, the results may perhaps surprise us; and if we have not done it before, may lead us to reflection. Let us see what some of them are.

The people of the United States are believed to eat, upon the average, an amount of animal food equal at least to one whole meal once a day, and those of Great Britain one in two days. But taking this estimate to be correct, Great Britain, by substituting vegetable for animal food, might sustain forty-nine instead of twenty-one millions of inhabitants, and the United States sixty-six millions instead of twenty; and this, too, in their present comfort, and without clearing up any more new land. Here, then, we are consuming that unnecessarily—if animal food is unnecessary—which would sustain seventy-nine millions of human beings in life, health, and happiness.

Now, if life is a blessing at all—if it is a blessing to twenty-two millions in Great Britain, and twenty millions in the United States—then to add to this population an increase of seventy-nine millions, would be to increase, in the same proportion, the aggregate of human happiness. And if, in addition to this, we admit the very generally received principle, that there is a tendency, from the nature of things, in the population of any country, to keep up with the means of support, we, of Great Britain and America, keep down, at the present moment, by flesh-eating, sixty-three millions of inhabitants.

We do not destroy them, in the full sense of the term, it is true, for they never had an existence. But we prevent their coming into the possession of a joyous and happy existence; and though we have no name for it, is it not a crime? What! no crime for thirty-five millions of people to prevent and preclude the existence of sixty-three millions?

I see no way of avoiding the force of this argument, except by denying the premises on which I have founded my conclusions. But they are far more easily denied than disproved. The probability, after all, is, that my estimates are too low, and that the advantages of an exclusively vegetable diet, in a national or political point of view, are even greater than is here represented. I do not deny, that some deduction ought to be made on account of the consumption of fish, which does not prevent the growth or use of vegetable products; but my belief is, that, including them, the animal food we use amounts to a great deal more than one meal a day, or one third of our whole living.

Suppose there was no *crime* in shutting human beings out of existence by flesh-eating, at the amazing rate I have mentioned—still, is it not, I repeat it, a great national or political loss? Or, will it be said, in its defence, as has been said in defence of war, if not of intemperance and some of the forms of licentiousness, that as the world is, it is a blessing to keep down its population, otherwise it would soon be overstocked? The argument would be as good in one case as in the other; that is, it is not valid in either. The world might be made to sustain, in comfort, even in the present comparatively infant state of the arts and sciences, at least forty or fifty times its present number of inhabitants. It will be time enough a thousand or two thousand years to come, to begin to talk about the danger of the world's being over-peopled; and, above all, to talk about justifying what we know is, in the abstract, very wrong, to prevent a distant imagined evil; one, in fact, which may not, and probably will not ever exist.

V. THE ECONOMICAL ARGUMENT.

The economy of the vegetable system is so intimately connected with its political or national advantages; that is, so depends on, or grows out of them, that I hesitated for some time before I decided to consider it separately. Whatever is shown clearly to be for the general good policy and well-being of society, cannot be prejudicial to the best interests of the individuals who compose that society. Still, there are some minor considerations that I wish to present under this head, that could not so well have been introduced any where else.

There is, indeed, one reason for omitting wholly the consideration of the pecuniary advantages of the system which I am attempting to defend. The public, to some extent, at once consider him who adverts to this topic, as parsimonious or mean. But, conscious as I am of higher objects in consulting economy than the saving of money, that it may be expended on things of no more value than the mere indulgence or gratification of the appetites or the passions, in a world where there are minds to educate and souls to save, I have ventured to treat on the subject.

It must be obvious, at a single glance, that if the vegetable products of an acre of land—such as wheat, rye, corn, barley, potatoes, beans, peas, turnips, beets, apples, strawberries, etc.—will sustain a family in equal health eight times as long as the pork, or beef, or mutton, which the same vegetables would make by feeding them to domestic animals, it must be just as mistaken a policy for the individual to make the latter disposition of these products as for a nation to do so. Nations are made of individuals; and, as I have already said, whatever is best, in the end, for the one, must also be the best, as a general rule, for the other.

But who has not been familiar from his very infancy with the maxim, that "a good garden will half support a family?" And who that is at all informed in regard to the manners and customs of the old world, does not know that the maxim has been verified there, time immemorial? But again: who has not considered, that if a garden of a given size will half support a family, one twice as large would support it wholly?

The truth is, it needs but a very small spot indeed, of good soil, for raising all the necessaries of a family. I think I have shown, in another work,[23] that five hundred and fifty pounds of Indian or corn meal, or ten bushels of the corn, properly cooked, will support, or more than support, an adult individual a year. Four times this amount is a very large allowance for a family of five persons; nay, even three times is sufficient. But how small a spot of good soil is required for raising thirty bushels of corn!

It is true, no family would wish to be confined a whole year to this one kind of food; nor do I wish to have it so; not that I think any serious mischiefs would arise as the consequence; but I should prefer, for my own part, a greater variety. But this does not materially alter the case. Suppose an acre and a half of land were required for the production of thirty bushels of corn. Let the cultivator, if he chooses, raise only fifteen bushels of corn, and sow the remainder with barley, or rye, or wheat. Or, if he prefer it, let him plant the one half of the piece with beans, peas, potatoes, beets, onions, etc. The one half of the space devoted to the production of some sort of grain would still half support his family; and it would require more than ordinary gluttony in a family of five persons to consume the produce of the other half, if the crops were but moderately abundant. A quarter of an acre of it ought to produce, at least, sixty bushels of potatoes; but this alone, would give such a family about ten pounds of potatoes, or one sixth of a bushel a day, for every day in the year, which is a tolerable allowance of food, without the grain and other vegetables.

But suppose a whole family were to live wholly on grain, as corn, or even wheat, for the year; the whole expenditure would hardly, exceed fifty dollars, in dear places and in the dearest times. Of course, I am speaking now of expenses for food and drink merely, the latter of which usually costs nothing, or need not. How small a sum is this to expend in New York, or Boston, or Philadelphia, in the maintenance of a family! And yet, it is amply sufficient for the vegetable-eater, unless his family live exclusively on wheat bread, or milk, when it might fall a little short. Of corn, at a dollar a bushel, it would give him eight pounds a day—far more than a family ought to consume, if they ate nothing else; and of potatoes, at forty cents a bushel, above twenty pounds, or one third of a bushel—more than sufficient for the family of an Hibernian.

Now, let me ask how much beef, or lamb, or pork, or sausages, or eggs, or cheese, this would buy? At ten cents a pound for each, which is comparatively low, it would buy five hundred pounds; about one pound and six ounces for the whole family, or four or five ounces each a day. This would be an average amount of nutriment equal to that of about two ounces of grain, or bread of grain, a day, to each individual. In so far as laid out in butter, or chicken, or turkey, at twenty cents a pound, it would give also about two or three ounces a day!

Further remarks under this head can hardly be necessary. He who considers the subject in its various aspects, will be likely to see the weight of the argument. There is a wide difference between a system which will give to each member of a family, upon the average, only about four or five ounces of food a day, and one which will give each of them more than twenty-five ounces a day, each ounce of the latter containing twice the nutriment of the former, and being much more savory and healthy at the same time. There is a wide difference, in matters of economy, at least, between ONE and TEN.

I will only add, under this head, a few tables. The first is to show the comparative amount of nutritious matter contained in some of the leading articles of human food, both animal and vegetable. It is derived from the researches of such men as MM. Percy and Vauquelin, of France, and Sir Humphrey Davy, of England.

100	pounds of	Wheat	contain	85	pounds	of	nutritious matter.
"	"	Rice	"	90	"	"	"
"	"	Rye	"	80	"	"	"
"	"	Barley	"	83	"	"	"
"	"	Peas	"	93	"	"	"
"	"	Lentils	"	94	"	"	"
"	"	Beans		89 to 92	"	"	"
"	"	Bread	(average)	80	"	"	"
"	"	Meat	(average)	35	"	"	"
"	"	Potatoes	contain	25	"	"	"

"	"	Beets	"	14	"	"	"
"	"	Carrots	"	10 to 14	"	"	"
"	"	Cabbage	"	7	"	"	"
"	"	Greens, turnips	"	4 to 8	"	"	"

Of course, it does not follow that every individual will be able to extract just this amount of nutriment from each article; for, in this respect, as well as in others, much will depend on circumstances.

The second table is from Mr. James Simpson, of Manchester, England, in a small work entitled, "The Products of the Vegetable Kingdom versus Animal Food," recently published in London. Its facts are derived from Dr. Playfair, Boussingault, and other high authorities. It will be seen to refute, entirely, the popular notions concerning the Liebig theory. The truth is, Liebig's views are misunderstood. His views are not so much opposed to mine as many suppose. Besides, neither he nor I are infallible.

	Solid matter.	Water.	Flesh forming principle.	Heat forming principle.	Ashes for the bones.
Potatoes,	28 per ct.	72 per ct.	2 per ct.	25 per ct.	1 per ct.
Turnips,	11 "	89 "	1 "	9 "	1 "
Barley Meal,	84-1/2 "	15-1/2 "	14 "	68-1/2 "	2 "
Beans,	86 "	14 "	31 "	51-1/2 "	3 "
Oats,	82 "	18 "	11 "	68 "	3 "
Wheat,	85-1/2 "	14-1/2 "	21 "	62 "	2-1/2 "

Peas,	84	"	16	"	29	"	51-1/2	"	3-1/2	"
Carrots,	13	"	87	"	2	"	10	"	1	"
Veal,	25	"	75	"	{					
Beef,	25	"	75	"	{25					
Mutton,	25	"	75	"	{					
Lamb,	25	"	75	"	{					
Blood,	20	"	80	"	20					

VI. THE ARGUMENT FROM EXPERIENCE.

A person trained in the United States or in England—but especially one who was trained in New England—might very naturally suppose that all the world were flesh-eaters; and that the person who abstains from an article which is at almost every one's table, was quite singular. He would, perhaps, suppose there must be something peculiar in his structure, to enable him to live without either flesh or fish; particularly, if he were a laborer. Little would he dream—little does a person who has not had much opportunity for reading, and who has not been taught to reflect, and who has never traveled a day's journey from the place which gave him birth, even so much as dream—that almost all the world, or at least almost all the hard-laboring part of it, are vegetable-eaters, and always have been; and that it is only in a few comparatively small portions of the civilized and half-civilized world, that the bone and sinew of our race ever eat flesh or fish for any thing more than as a condiment or seasoning to the rest of their food, or even taste it at all. And yet such is the fact.

It is true, that in a vast majority of cases, as I have already intimated, laborers are vegetable-eaters from necessity: they cannot get flesh. Almost all mankind, as they are usually trained, are fond of extra stimulants, if they can get them; and whether they are called savages or civilized men, will indulge in them more or less, if they are to be had, unless their intellectual and moral natures have been so well developed and cultivated, as to have acquired the ascendency. Spirits, wine, cider, beer, coffee, tea, condiments, tobacco, opium, snuff, flesh meat, and a thousand other things, which excite, for a time, more pleasurable sensations than water and plain vegetables and fruits, will be sought with more or less eagerness according to the education which has been received, and according to our power of self-government.

I have said that most persons are vegetable-eaters from necessity, not from choice. There are some tribes in the equatorial regions who seem to be exceptions to this rule; and yet I am not quite satisfied they are so. Some children, among us, who are trained to a very simple diet, will seem to shrink from tea or coffee, or alcohol, or camphor, and even from any thing which is much heated, when first presented to them. But, train the same children to the ordinary, complex, high-seasoned diet of this country, and it will not take long to find out that they are ready to acquire the habit of relishing the excitement of almost all sorts of *unnaturals* which can be presented to them. And if there are tribes of men who at first refuse flesh meat, I apprehend they do so for the same reasons which lead a child among us, who is trained simply to refuse hot food and drink, or at least, hot tea and coffee, when the latter are first presented to him.

Gutzlaff, the Chinese traveler and missionary, has found that the Chinese of the interior, who have scarcely ever tasted flesh or fish, soon acquire a wonderful relish for it, just as our children do for spirituous or exciting drinks and drugs, and as savages do for tobacco and spirits. But he has also made another discovery, which is, that flesh-eating almost ruins them for labor. Instead of being strong, robust, and active, they soon become lazy, self-indulgent, and effeminate. This is a specimen—perhaps a tolerably fair one—of the natural tendency of such food in all ages and countries. Man every where does best, nationally and individually, other things being equal, on a well-chosen diet of vegetables, fruits, and water. In proportion as individuals or families, or tribes or nations, depart from this—other things being equal—in the same proportion do they degenerate physically, intellectually, and morally.

Such a statement may startle some of my New England readers, perhaps, who have never had opportunity to become acquainted with facts as they are. But can it be successfully controverted? Is it not true, that, with a few exceptions—and those more apparent than real—nations have flourished, and continued to flourish, in proportion as they have retained the more natural dietetic habits to which I have alluded; and that they have been unhappy or short-lived, as nations, in proportion as exciting food and drink have been used? Is it not true, that those individuals, families, tribes, and nations, which have used what I call excitements, liquid or solid, have been subjected by them to the same effects which follow the use of spirits—first, invigoration, and subsequently decline, and ultimately a loss of strength? Why is it that the more wealthy, all over Europe, who get flesh more or less, deteriorate in their families so rapidly? Why is it that every thing is, in this respect, so stationary among the middle classes and the poor?

In short—for the case appears to me a plain one—it is the simple habits of some, whether we speak of nations, families, or individuals, which have preserved the world from going to utter decay. In ancient times, the Egyptians, the most enlightened and one of the most enduring of nations, were what might properly be called a vegetable-eating nation; so were the ancient Persians, in the days of their greatest glory; so the Essenes, among the Jews; so the Romans, as I have said elsewhere, and the Greeks. If either Moses or Herodotus is to be credited, men lived, in ancient times, about a thousand years. Indeed, empire seems to have departed from among the ancient nations precisely when simplicity departed. So it is with nations still. A flesh-eating nation may retain the supremacy of the world a short time, as several European and American nations have done; just as the laborer, whose brain and nerves are stimulated by ardent spirits, may for a time retain—through the medium of an artificial strength—the ascendency among his fellow-laborers; but the triumph of both the nation and the individual must be short, and the debility which follows proportionable. And if the United States, as a nation, seem to form an exception to the truth of this remark, it is only because the stage of debility has not yet arrived. Let us be patient, however, for it is not far off.

But to come to the specification of facts. The Japanese of the interior, according to some of the British geographers, live principally on rice and fruits—a single handful of rice often forming the basis of their frugal meal. Flesh, it is said, they either cannot get, or do not like; and to milk, even, they have the same sort of aversion which most of us have to blood. It is only a few of them, comparatively, and those principally who live about the coasts, who ever use either flesh or fish. And yet we have the concurring testimony of all geographers and travelers, that in their physical and intellectual development, at least, to say nothing of their moral peculiarities, they are the finest men in all Asia. In what country of Asia are schools and early education in such high reputation as in Japan? Where are the inhabitants so well formed, so stout made, and so robust? Compare them with the natives of New Holland, in the same, or nearly the same longitude, and about as far south of the equator as the Japanese are north of it, and what a contrast! The New Hollanders, though eating flesh liberally, are not only mere savages, but they are among the most meagre and wretched of the human race. On the contrary, the Japanese, in mind and body, are scarcely behind the middle nations of Europe.

Nearly the same remarks will apply to China, and with little modification, to Hindostan. In short, the hundreds of millions of southern Asia are, for the most part, vegetable-eaters; and a large proportion of them live chiefly, if not wholly on rice, though by no means the most favorable vegetable for exclusive use. What countries like these have maintained their ancient, moral, intellectual, and political landmarks? Grant that they have made but little improvement from century to century; it is something not to have deteriorated. Let us proceed with our general view of the world, ancient and modern.

The Jews of Palestine, two thousand years ago, lived chiefly on vegetable food. Flesh, of certain kinds, was indeed admissible, by their law; but, except at their feasts and on special occasions, they ate chiefly bread, milk, honey, and fruits.

Lawrence says that "the Greeks and Romans, in the periods of their greatest simplicity, manliness, and bravery, appear to have lived almost entirely on plain vegetable preparations."

The Irish of modern days, as well as the Scotch, are confined almost wholly to vegetable food. So are the Italians, the Germans, and many other nations of modern Europe. Yet, where shall we look for finer specimens of bodily health, strength, and vigor, than in these very countries? The females, especially, where shall we look for their equals? The men, even—the Scotch and Irish, for example—are they weaker than their brethren, the English, who use more animal food?

It will be said, perhaps, the vegetable-eating Europeans are not always distinguished for vigorous minds. True; but this, it may be maintained, arises from their degraded physical condition, generally; and that neglect of mental and moral cultivation which accompanies it. A few, even here, like comets in the material system, have occasionally broken out, and emitted no faint light in the sphere in which they were destined to move.

But we are not confined to Europe. The South Sea Islanders, in many instances, feed almost wholly on vegetable substances; yet their agility and strength are so great, that it is said "the stoutest and most expert English sailors, had no chance with them in wrestling and boxing."

We come, lastly, to Africa, the greater part of whose millions feed on rice, dates, etc.; yet their bodily powers are well known.

In short, more than half of the 800,000,000 of human beings which inhabit our globe live on vegetables; or, if they get meat at all, it is so rarely that it can hardly have any effect on their structure or character. Out of Europe and the United States—I might even say, out of the latter—the use of animal food is either confined to a few meagre, weak, timid nations, like the Esquimaux, the Greenlanders, the Laplanders, the Samoiedes, the Kamtschadales, the Ostiacs, and the natives of Siberia and Terra del Fuego; or those wealthier classes, or individuals of every country, who are able to range lawlessly over the Creator's domains, and select, for their tables, whatever fancy or fashion, or a capricious appetite may dictate, or physical power afford them.

VII. THE MORAL ARGUMENT.

In one point of view, nearly every argument which can be brought to show the superiority of a vegetable diet over one that includes flesh or fish, is a moral argument.

Thus, if man is so constituted by his structure, and by the laws of his animal economy, that all the functions of the body, and of course all the faculties of the mind, and the affections of the soul, are in better condition—better subserve our own purposes, and the purposes of the great Creator—as well as hold out longer, on the vegetable system—then is it desirable, in a moral point of view, to adopt it. If mankind lose, upon the average, about two years of their lives by sickness, as some have estimated it,[24] saying nothing of the pain and suffering undergone, or of the mental anguish and soul torment which grow out of it, and often render life a burden; and if the simple primitive custom of living on vegetables and fruits, along with other good physical and mental habits, which seem naturally connected with it, will, in time, nearly if not wholly remove or prevent this amazing loss, then is the argument deduced therefrom, in another part of this chapter, a moral argument.

If, as I have endeavored to show, the adoption of the vegetable system by nations and individuals, would greatly advance the happiness of all, in every known respect, and if, on this account, such a change in our flesh-eating countries would be sound policy, and good economy,—then we have another moral argument in its favor.

But, again; if it be true that all nations have been the most virtuous and flourishing, other things being equal, in the days of their simplicity in regard to food, drink, etc.; and if we can, in every instance, connect the decline of a nation with the period of their departure, as a nation, into the maze of luxurious and enervating habits; and if this doctrine is, as a general rule, obviously applicable to smaller classes of men, down to single families, then is the argument we derive from it in its nature a moral one. Whatever really tends, without the possibility of mistake, to the promotion of human happiness, here and hereafter, is, without doubt, moral.

But this, though much, is not all. The destruction of animals for food, in its details and tendencies, involves so much of cruelty as to cause every reflecting individual—not destitute of the ordinary sensibilities of our nature—to shudder. I recall: daily observation shows that such is not the fact; nor should it, upon second thought, be expected. Where all are dark, the color is not perceived; and so universally are the moral sensibilities which really belong to human nature deadened by the customs which prevail among us, that few, if any, know how to estimate, rightly, the evil of which I speak. They have no more a correct idea of a true sensibility—not a *morbid* one—on this subject, than a blind man has of colors; and for nearly the same reasons. And on this account it is, that I seem to shrink from presenting, at this time, those considerations which, I know, cannot, from the very nature of the case, be properly understood or appreciated, except by a very few.

Still there are some things which, I trust, may be made plain. It must be obvious that the custom of rendering children familiar with the taking away of life, even when it is done with a good degree of tenderness, cannot have a very happy effect. But, when this is done, not only without tenderness or sympathy, but often with manifestations of great pleasure, and when children, as in some cases, are almost constant witnesses of such scenes, how dreadful must be the results!

In this view, the world, I mean our own portion of it, sometimes seems to me like one mighty slaughter-house—one grand school for the suppression of every kind, and tender, and brotherly feeling—one grand process of education to the entire destitution of all moral principle—one vast scene of destruction to all moral sensibility, and all sympathy with the woes of those around us. Is it not so?

I have seen many boys who shuddered, at first, at the thought of taking the life, even of a snake, until compelled to it by what they conceived to be duty; and who shuddered still more at taking the life of a lamb, a calf, a pig, or a fowl. And yet I have seen these same boys, in subsequent life, become so changed, that they could look on such scenes not merely with indifference, but with gratification. Is this change of feeling desirable? How long is it after we begin to look with indifference on pain and suffering in brutes, before we begin to be less affected than before by human suffering?

I am not ignorant that sentiments like these are either regarded as morbid, and therefore pitiable, or as affected, and therefore ridiculous. Who that has read the story of Anthony Benezet, as related by Dr. Rush, has not smiled at what he must have regarded a feeling wholly misplaced, if nothing more? And yet it was a feeling which I think is very far from deserving ridicule, however homely the manner of expressing it. But I have related this interesting story in another part of the work.

I am not prepared to maintain, strongly, the old-fashioned doctrine, that a butcher who commences his employment at adult age, is necessarily rendered hardhearted or unfeeling; or, that they who eat flesh have their sensibilities deadened, and their passions inflamed by it—though I am not sure that there is not some truth in it. I only maintain, that to render children familiar with the taking away of animal life,—especially the lives of our own domestic animals, often endeared to us by many interesting circumstances of their history, or of our own, in relation to them,—cannot be otherwise than unhappy in its tendency.

How shocking it must be to the inhabitants of Jupiter, or some other planet, who had never before witnessed these sad effects of the ingress of sin among us, to see the carcasses of animals, either whole or by piece-meal, hoisted upon our very tables before the faces of children of all ages, from the infant at the breast, to the child of ten or twelve, or fourteen, and carved, and swallowed; and this not merely once, but from day to day, through life! What could they—what would they—expect from such an education of the young mind and heart? What, indeed, but mourning, desolation, and woe!

On this subject the First Annual Report of the American Physiological Society thus remarks—and I wish the remark might have its due weight on the mind of the reader:

"How can it be right to be instrumental in so much unnecessary slaughter? How can it be right, especially for a country of vegetable abundance like ours, to give daily employment to twenty thousand or thirty thousand butchers? How can it be right to train our children to behold such slaughter? How can it be right to blunt the edge of their moral sensibilities, by placing before them, at almost every meal, the mangled corpses of the slain; and not only placing them there, but rejoicing while we feast upon them?"

One striking evidence of the tendency which an habitual shedding of blood has on the mind and heart, is found in the fact that females are generally so reluctant to take away life, that notwithstanding they are trained to a fondness for all sorts of animal food, very few are willing to gratify their desires for a stimulating diet, by becoming their own butchers. I have indeed seen females who would kill a fowl or a lamb rather than go without it; but they are exceedingly rare. And who would not regard female character as tarnished by a familiarity with such scenes as those to which I have referred? But if the keen edge of female delicacy and sensibility would be blunted by scenes of bloodshed, are not the moral sensibilities of our own sex affected in a similar way? And must it not, then, have a deteriorating tendency?

It cannot be otherwise than that the circumstances of which I have spoken, which so universally surround infancy and childhood, should take off, gradually, the keen edge of moral sensibility, and lessen every virtuous or holy sympathy. I have watched—I believe impartially—the effect on certain sensitive young persons in the circle of my acquaintance. I have watched myself. The result has confirmed the opinion I have just expressed. No child, I think, can walk through a common market or slaughter-house without receiving moral injury; nor am I quite sure that any virtuous adult can.

How have I been struck with the change produced in the young mind by that merriment which often accompanies the slaughter of an innocent fowl, or lamb, or pig! How can the Christian, with the Bible in hand, and the merciful doctrines of its pages for his text,

"Teach me to feel another's woe,"

—the beast's not excepted—and yet, having laid down that Bible, go at once from the domestic altar to make light of the convulsions and exit of a poor domestic animal?

Is it said, that these remarks apply only to the *abuse* of a thing, which, in its place, is proper? Is it said, that there is no necessity of levity on these occasions? Grant that there is none; still the result is almost inevitable. But there is, in any event, one way of avoiding, or rather preventing both the abuse and the occasion for abuse, by ceasing to kill animals for food; and I venture to predict that the evil never will be prevented otherwise.

The usual apology for hunting and fishing, in all their various and often cruel forms,—whereby so many of our youth, from the setters of snares for birds, and the anglers for trout, to the whalemen, are educated to cruelty, and steeled to every virtuous and holy sympathy,—is, the necessity of the animals whom we pursue for food. I know, indeed, that this is not, in most cases, the true reason, but it is the reason given—it is the substance of the reason. It serves as an apology. They who make it may often be ignorant of the true reason, or they or others may wish to conceal it; and, true to human nature, they are ready to give every reason for their conduct, but the real and most efficient one.

It must not, indeed, be concealed that there is one more apology usually made for these cruel sports; and made too, in some instances, by good men; I mean, by men whose intentions are in the main pure and excellent. These sports are healthy, they tell us. They are a relief to mind and body. Perhaps no good man, in our own country, has defended them with more ingenuity, or with more show of reason and good sense, than Dr. Comstock, in his recent popular work on Human Physiology. And yet, there is scarcely a single advantage which he has pointed out, as being derived from the "pleasures of the chase," that may not be gained in a way which savors less of blood. The doctor himself is too much in love with botany, geology, mineralogy, and the various branches of natural history, not to know what I mean when I say this. He knows full well the excitement, and, on his own principles, the consequent relief of body and mind from their accustomed and often painful round, which grows out of clambering over mountains and hills, and fording streams, and climbing trees and rocks, to need any very broad hints on the subject; to say nothing of the delights of agriculture and horticulture. How could he, then, give currency to practices which, to say the least,—and by his own concessions, too,—are doubtful in regard to their moral tendencies, by inserting his opinions in favor of sports, for which he himself happens to be partial, in a school-book? Is this worthy of those who would educate the youth of our land on the principles of the Bible?

VIII. THE MILLENNIAL ARGUMENT

I believe it is conceded by most intelligent men, that all the arguments we bring against the use of animal food, which are derived from anatomy, physiology, or the laws of health, or even of psychology, are well founded. But they still say, "Man is not what he once was; he is strangely perverted; that custom, or habit, which soon becomes second nature, and often proves stronger to us than first nature, has so changed him that he is more a creature of art than of nature, or at least of *first* nature. And though animal food was not necessary to him at first—perhaps not in accordance with his best interests—yet it has become so by long use; and as a creature of art rather than of nature, he now seems to require it."

This reasoning, at first view, appears very *specious*. But upon second view, we see it is wanting—greatly so—in solidity. It takes for granted, as I understand it, that what we call civilization, has rendered animal food necessary to man. But is it not obvious that the condition of things which is thus supposed to render this species of food necessary, is not likely to disappear—nay, that it is every century becoming more and more the law, so to speak, of the land? Who is to stop the labor-saving machine, the railroad car, or the lightning flash of intelligence?

And do not these considerations, if they prove any thing, prove quite too much? For if, in the onward career of what is thus called civilization, we have gone from a diet which scarcely required the use of animal food in order to render it both palatable and healthful, to one in whose dishes it is generally blended in some one or more of its forms, must we not expect that a still further progress in the same course will render the same kind of diet still more indispensable? If flesh, fish, fowl, butter, cheese, eggs, lard, etc., are much more necessary to us now, than they were a thousand years ago, will they not be still more necessary a thousand years hence?

I do not see how we can avoid such a conclusion. And yet such a conclusion will involve us in very serious difficulties. In Japan and China—the former more especially—if the march of civilization should be found to have rendered animal food more necessary, it has at the same time rendered it less accessible to the mass of the population. The great increase of the human species has crowded out the animals, even the domestic ones. Some of the old historians and geographers tell us that there are not so many domestic animals in the whole kingdom of Japan, as in a single township of Sweden.

And must not all nations, as society progresses and the millennium dawns, crowd out the animals in the same way? It cannot be otherwise. True, there may remain about the same supply as at present from the rivers and seas, and perchance from the air; but what can these do for the increasing hundreds of millions of such large countries? What do they for Japan? In short, if the reasoning above were good and valid, it would seem to show that precisely at the point of civilization where animal food becomes most necessary, at precisely that point it becomes most scarce.

These things do not seem to me to go well together. We must reject the one or the other. If we believe in a millennium, we must, inevitably, give up our belief in animal food, at least the belief that its necessity grows out of the increasing wants of society. Or if, on the other hand, we believe in the increasing necessity of animal food, we must banish from our minds all hope of what we call a millennium, at least for the present.

IX. THE BIBLE ARGUMENT.

It is not at all uncommon for those who find themselves driven from all their strong-holds, in this matter, to fly to the Bible. Our Saviour ate flesh and fish, say they; and the God of the New Testament, as well as of the Old, in this and other ways, not only permitted but sanctioned its use.

But, to say nothing of the folly of going, for proof of every thing we wish to prove, to a book which was never given for this purpose, or of the fact that in thus adducing Scripture to prove our favorite doctrines, we often go too far, and prove too much; is it true that the Saviour ate flesh and fish? Or, if this could be proved, is it true that his example binds us forever to that which other evidence as well as science show to be of doubtful utility? Paul did not think so, most certainly. It is good neither to eat flesh nor to drink wine, he says, if it cause our brother to offend. Did not Paul understand, at least as well as we, the precepts and example of our Saviour?

And as to a permission to Noah and his descendants, the Jews, to use animal food—was it not for the hardness of the human heart, as our Saviour calls it? From the beginning, was it so? Is not man, in the first chapter of Genesis, constituted a vegetable-eater? Was his constitution ever altered? And if so, when and where? Will they who fly to the Bible for their support, in this particular, please to tell us?

But it is idle to go to the Bible, on this subject. I mean, it is idle to pretend to do so, when we mean not so much. Men who *incline* to wine and other alcoholic drinks, plead the example and authority of the Bible. Yet you will hardly find a man who drinks wine simply because he believes the Bible justifies its use. He drinks it for other reasons, and then makes the foolish excuse that the Bible is on his side. So in regard to the use of flesh meat. Find a man who really uses flesh or fish *because* the Bible requires him to do so, and I will then discuss the question with him on Bible ground. Till that time, further argument on this direction is unnecessary.

CONCLUSION.

But I must conclude this long essay. There is one consideration, however, which I am unwilling to omit, although, in deciding on the merits of the question before us, it may not have as much weight—regarded as a part of the moral argument—on every mind, as it has on my own.

Suppose the great Creator were to make a new world somewhere in the regions of infinite space, and to fit it out in most respects like our own. It is to be the place and abode of such minerals, vegetables, and animals as our own. Instead, however, of peopling it gradually, he fills it at once with inhabitants; and instead of having the arts and the sciences in their infancy, he creates every thing in full maturity. In a word, he makes a world which shall be exactly a copy of our own, with the single exception that the 800,000,000 of free agents in it shall be supposed to be wholly ignorant in regard to the nature of the food assigned them. But the new world is created, we will suppose, at sunrise, in October. The human inhabitants thereof have stomachs, and soon, that is, by mid-day or before night, feel the pangs of hunger. Now, what will they eat?

The world being mature, every thing in it is, of course, mature. Around, on every hand, are cornfields with their rich treasures; above, that is, in the boughs of the orchards, hang the rich russets, pippins, and the various other excellent kinds of the apple, with which our own country and other temperate climates abound. In tropical regions, of course, almost every vegetable production is flourishing at that season, as well as the corn and the apple. Or, he has but to look on the surface of the earth on which he stands, and there are the potatoe, the turnip, the beet, and many other esculent roots; to say nothing of the squash, the pumpkin, the melon, the chestnut, the walnut, the beechnut, the butternut, the hazelnut, etc.,—most of which are nourishing, and more or less wholesome, and are in full view. Around him, too, are the animals. I am willing even to admit the domestic animal—the horse, the ox, the sheep, the dog, the cat, the rabbit, the turkey, the goose, the hen, yes, and even the pig. And now, I ask again, what will he eat? He is destitute of experience, and he has no example. But he has a stomach, and he is hungry: he has hands and he has teeth; the world is all before him, and he is the lord of it, at least so far as to use such food in it as he pleases.

Does any one believe that, in these circumstances, man would prey upon the animals around him? Does any person believe—can he for one moment believe—he would forthwith imbrue his hands in blood, whether that of his own species or of some other? Would he pass by the mellow apple, hanging in richest profusion every where, inviting him as it were by its beauties? Would he pass by the fields, with their golden ears? Would he despise the rich products of field, and forest, and garden, and hasten to seize the axe or the knife, and, ere the blood had ceased to flow, or the muscles to quiver, give orders to his fair but affrighted companion within to prepare the fire, and make ready the gridiron or the spider? Or, without the knowledge even of this, or the patience to wait for the tedious process of cooking to be completed, would he eat raw the precious morsel? Does any one believe this? Can any one—I repeat the question—can any one believe it?

On the contrary, would not every living human being revolt, at first, from the idea, let it be suggested as it might, of plunging his hands in blood? Can there be a doubt that he would direct his attention at first—yes, and for a long time afterward—to the vegetable world for his food? Would it not take months and years to reconcile his feelings—his moral nature—to the thought of flesh-mangling or flesh-eating? At least, would not this be the result, if he were a disciple of Christianity? Although professing Christians, as the world is now constituted, do not hesitate to commit such depredations, would they do so in the circumstances we have supposed?

I am sure there can be but one opinion on this subject; although I confess it impossible for me to say how it may strike other minds constituted somewhat differently from my own. With me, this consideration of the subject has weight and importance. It is not necessary, however. The argument—the moral argument, I mean—is sufficient, as it seems to me, without it. What then shall we say of the anatomical, the physiological, the medical, the political, the economical, the experimental, the Bible, the millennial, and the moral arguments, when united? Have they not force? Are they not a ninefold cord, not easily broken? Is it not too late in the day of human improvement to meet them with no argument but ignorance, and with no other weapon but ridicule?

FOOTNOTES:

[21] For proof that arsenic or ratsbane is sometimes added to cheese, see the Library of Health, volume ii., page 69. In proof of the poisonous tendency of milk and butter, see Whitlaw's Theory of Fever, and Clark's Treatise on Pulmonary Consumption.

[22] See Dunglison's Hygiene, page 250.

[23] The Young Housekeeper.

[24] Or, more nearly, perhaps, a year and a half, in this country. In England, it is one year and five-sevenths.

OUTLINES OF A
NEW SYSTEM OF FOOD AND COOKERY.

In the work of revising and preparing the foregoing volume for publication, the writer was requested to add to it a system of vegetable cookery. At first he refused to do so, both on account of the difficulty of bringing so extensive a subject within the compass of twenty or thirty pages, and because it did not seem to him to be called for, in connection with the present volume. But he has yielded his own judgment to the importunity of the publishers and other friends of the work, and prepared a mere outline or skeleton of what he may hereafter fill up, should circumstances and the necessary leisure permit.

But there is one difficulty to be met with at the very threshold of the subject. Vegetable eaters are not so hard driven to find whereon to subsist, as many appear to suppose. For the question is continually asked, "If you dispense wholly with flesh and fish, pray what can you find to eat?" Now, while we are aware that one small sect of the vegetarians—the followers of Dr. Schlemmer—eat every thing in a raw state, we are, for ourselves, full believers in plain and simple cookery. That a potato, for example, is better cooked than uncooked, both for man and beast, we have not the slightest doubt. We believe that a system of preparing food which renders the raw material more palatable, more digestible, and more nutritious, or perhaps all this at once, must be legitimate, and even preferable—if not for the individual, at least for the race.

But the difficulty alluded to is, how to select a few choice dishes from the wide range—short of flesh and fish—which God and nature permit. For if we believed in the use of eggs when commingled with food, we should hardly deem it proper to go the whole length of our French brethren, who have nearly seven hundred vegetable dishes, of which eggs form a component part; nor the whole length even to which our own powers of invention might carry us; no, nor even the whole length to which the writer of an English work now before us, and entitled "Vegetable Cookery," has gone—the extent of about a thousand plain receipts. We believe the whole nature of man, and even his appetite, when unperverted, is best served and most fully satisfied with a range of dishes which shall hardly exceed hundreds.

It is held by Dr. Dunglison, Dr. Paris, and many of the old school writers, that all made dishes—all mixtures of food—are "more or less rebellious;" that is, more or less indigestible, and consequently more or less hurtful. If they mean by this, that in spite of the accommodating power of the stomach to the individual, they are hurtful to the race, I go with them most fully. But I do *not* believe that *all made dishes, to all persons*, are so directly injurious as many suppose. God has made man, in a certain sense, omnivorous. His physical stomach can receive and assimilate, like his mental stomach, a great variety of substances; and both can go on, without apparent disease, for a great many years, and perhaps for a tolerably long life in this way.

There is, however, a higher question for man to ask as a rational being and as a Christian, than whether this or that dish will hurt him directly. It is, whether a dish or article is *best* for him—best for body, mind, and heart—best for the whole human nature—best for the whole interests of the whole race—best for time, and best for eternity. Startle not, reader, at this assertion. If West could properly say, "I paint for eternity," the true disciple of Christ and truth can say, "I eat and drink for eternity." And a higher authority than any that is merely human has even required us to do so.

This places the subject of preparing food on high ground. And were I to carry out my plan fully, I should exclude from a Christian system of food and cookery all mixtures, properly so called, and all medicines or condiments. Not that all mixtures are equally hurtful to the well-being of the race, nor all medicines. Indeed, considering our training and habits, some of both, to most persons, have become necessary. I know of many whose physical inheritance is such, that salt, if not a few other medicinal substances, have become at least present necessaries to them. And to those mixtures of substances closely allied, as farina with farina—meal of one kind with meal of another—I could scarcely have any objection, myself. Nature objects to incompatibles, and therefore I do; and medicine, and all those kinds of food which are opposed one to another, are incompatible with each other. When one is in the stomach, the other should not be.

I have spoken of carrying out my plan, but this I cannot now fully do. It would not be borne, till, as Lord Bacon used to say, "some time be passed over." But, on the other hand, I am unwilling to give directions, as I did ten or twelve years ago, in my Young Housekeeper, such as shall pander to a perverted—most abominably perverted—public taste. Man is made for progress, and it is high time the public standard were raised in regard to food and cookery.

Although grains and fruits are the natural food of man, yet there are a variety of shapes in which the grains or farinacea may be presented to us; and there are a few substances fit for food which do not properly belong to either of these classes. I shall treat first of the different kinds of food prepared from grain or farinaceous substances; secondly, of fruits; thirdly, of roots; and fourthly, speak of a few articles that do not properly belong to any of the three.

While, therefore, as will be seen by the remarks already made, I have many things to say that the community cannot yet bear, it need not escape the observation of the most careless reader, that I aim at nothing less than an entire ultimate subversion of the present system of cookery, believing it to be utterly at war with the laws of God, and of man's whole nature.

CLASS I.—FARINACEOUS, OR MEALY SUBSTANCES.

The principal of these are wheat, oats, Indian corn, rice, rye, barley, buckwheat, millet, chestnuts, peas, beans, and lentils. They are prepared in various forms.

DIVISION I.—BREAD.

The true idea of bread is that coarse or cracked and unbolted meal, formed into a mass of dough by means of water, and immediately baked in loaves of greater or less thickness, according to the fancy.

Some use bolted meal; most raise bread by fermentation; many use salt; some saleratus, or carbonate of potash; and, in the country, many use milk instead of water to form the paste. I might also mention several other additions, which, like saleratus, it is becoming fashionable to make.

All these things are a departure, greater or less, from the true idea of a bread; and bread made with any of these changes, is so much the less perfectly adapted to the promotion of health, happiness, and longevity.

Bolting is objectionable, because bread made from bolted meal, especially when eaten hot, is more apt, when the digestive powers are not very vigorous, to form a paste, which none but very strong stomachs can entirely overcome. Besides, it takes out a part of the sweetness, or life, as it is termed, of the flour. They who say fine flour bread is sweetest, are led into this mistake by the force of habit, and by the fact that the latter comes in contact, more readily than coarse bread, with the papillæ of the tongue, and seems to have more taste to it because it touches at more points.

Raising bread by inducing fermentation, wastes a part of the saccharine matter; and the more it is raised, the greater is the waste. By lessening the attraction of cohesion, it makes it more easy of digestion, it is true; but the loss of nutriment and of pleasure to the true appetite more than counterbalances this. Bakers, in striving to get a large loaf, rob the bread of most of its sweetness.

Salt is objectionable, because it hardens the bread, and renders it more difficult of digestion. Our ancestors, in this country, did not use it at all; and many are the families that will not use it now.

Those who use salt in bread, tell us how *flat* it would taste without it. This idea of flatness has two sources. 1. We have so long given our bread the taste of salt, as we have most other things, that it seems tasteless without it. 2. The flatness spoken of in an article of food is oftentimes the true taste of the article, unaltered by any stimulus. If any two articles need to be stimulated with salt, however, it is rice and beans—bread never.

If saleratus is used in bread where no acidity is present, it is a medicine; or, if you please, a poison both to the stomach and intestines. If it meets and neutralizes an acid either in the bread-tray or the stomach, the residuum is a new chemical compound diffused through the bread, which is more or less injurious, according to its nature and quantity.

Milk is objectionable on the score of its tendency to render the bread more indigestible than when it was wet with water, and perhaps by rendering it too nutritious. For good bread without the milk is already too nutritious for health, if eaten exclusively, for a long time. That man should not live on bread alone, is as true physically as it is morally.

No bread should be eaten while new and hot—though the finer it is, the worse for health when thus eaten. Old bread, heated again, is less hurtful. But if eaten both new and hot, and with butter or milk, or any thing which soaks and fills it, the effect is very bad. Mrs. Howland, in her Economical Housekeeper, says much about *ripe* bread. And I should be glad to say as much, had I room, about ripe bread, and about the true philosophy of bread and bread-making, as she has.

SECTION A.—*Bread of the first order.*

This is made of coarse meal—as coarse as it can well be ground, provided the kernels are all broken. The grain should be well washed, and it may be ground in the common way, or according to the oriental mode, in hand-mills. The latter mode is preferable, because you can thus have it fresh. Meal is somewhat injured by being kept long ground.

If great pains is not taken to have the grain clean when ground, it needs to be passed through a coarse sieve, that all foreign bodies may be carefully separated. The hulls of corn, and especially the husks of oats and buckwheat, should also be separated in some way. In no case, however, should meal be bolted. Good health requires that we eat the innutritious and coarser parts as well as the finer.

RECEIPT 1.—Take a sufficient quantity of good, recent wheat meal;[25] wet it well, but not too soft, with pure water; form it into thin cakes, and bake it as hard as the teeth will bear. Remember, however, that the saliva aids the teeth greatly, especially when you masticate your food slowly. The cakes should be very thin—the thinner the better. Many, however, prefer them an inch thick, or even more.

RECEIPT 2.—Oat meal prepared in the same manner. Procure what is called the Scotch kiln dried oat meal, if you can. No matter if it is manufactured in New England, if it is well done.

RECEIPT 3.—Indian meal cakes, otherwise called hoe cakes, or Johnny cakes, are next in point of value to bread made of wheat and oats. They are most healthy, however, in cold weather.

RECEIPT 4.—Rye cakes come next. Warm instead of cold water is often used to wet all the above. Some even choose to scald the meal. Fancy may be indulged in this particular, only you must remember that warm water in warm weather may soon give rise, if the mass stands long, to a degree of fermentation, which, for the best bread, should be avoided.

RECEIPT 5.—Barley meal bread comes next in order in the unleavened series. In regard to this species of bread, however, I do not speak from experience, but from report.

RECEIPT 6.—Of millet bread I know still less. Cakes made of it, as above, must certainly be wholesome.

RECEIPT 7.—Buckwheat cakes are last in the series of the best breads. The meal is always too fine, and hence makes heavy bread, except when hot. Few use it without fermentation.

Unleavened bread may be made as above, of all the various kinds of grain, finely ground; but it is apt to be heavy, whereas, when made properly, of coarse meal, it is only firm, never heavy; that is, it never has a lead-like appearance. They may make and use it who have iron stomachs.

SECTION B.—*Bread of the second order.*

This consists essentially of mixtures of the various coarse meals. True it is, that made or mixed food is objectionable; but the union of one farinaceous substance with another to form bread, can hardly be considered a mixture. It is, essentially, the addition of farina to farina, with some change in the proportion of the gluten and other properties.

RECEIPT 1.—Wheat meal and Indian, in about the proportion of two parts of wheat to one of Indian.

RECEIPT 2.—Wheat meal and oat meal, about equal parts.

RECEIPT 3.—Wheat meal and Indian, equal parts.

RECEIPT 4.—Wheat meal and rye meal; two parts, quarts, or pounds of the former to one of the latter.

RECEIPT 5.—Rye and Indian, equal parts of each.

RECEIPT 6.—Rye, two thirds; Indian, one third.

RECEIPT 7.—Wheat meal and rice. Three quarts of wheat meal to one pint of good clean rice, boiled till it is soft.

RECEIPT 8.—Three parts of wheat meal to one of Indian.

RECEIPT 9.—Four parts of wheat to one of Indian.

The proportion of the ingredients above may be varied to a great extent. I have inserted some of the best. The following are *irregulars*, but may as well be mentioned here as any where.

RECEIPT 10.—Two quarts of wheat meal to one pound of well boiled ripe beans, made soft by pounding or otherwise.

RECEIPT 11.—Seven pounds of wheat meal and two and a half pounds of good, mealy, and well boiled and pounded potatoes.

RECEIPT 12.—Equal parts of coarse meal from rye, barley, and buckwheat. This is chiefly used in Westphalia.

RECEIPT 13.—Seven parts of wheat meal (as in Receipt 11), with two pounds of split peas boiled to a soup, and used to wet the flour.

RECEIPT 14.—Wheat meal and apples, in the proportion of about three of the former (some use two) to one of the latter. The apples must be first pared and cored, and stewed or baked. See my "Young Housekeeper," seventh edition, page 396.

RECEIPT 15.—Wheat meal and boiled chestnuts; three quarts of the former to one of the latter.

RECEIPT 16.—Wheat meal, four quarts, and one quart of well boiled and pounded marrow squash.

RECEIPT 17.—Wheat, corn, or barley meal; three quarts to one quart of powdered comfrey root. This is inserted from the testimony of Rev. E. Rich, of Troy, N. H.

RECEIPT 18.—Wheat meal, three pounds, to one pound of pounded corn, boiled and pounded green. This is the most doubtful form which has yet been mentioned.

RECEIPT 19.—Receipt 7 describes rice bread. Bell, in his work on Diet and Regimen, says the best and most economical rice bread is made thus: Wheat meal, three pounds; rice, well boiled, one pound—wet with the water in which the rice is boiled.

I wish to say here, once for all, that any kind of bread may be salted, if you will *have* salt, except the patented bread mentioned in the beginning of the next section, which is salted in the process. Molasses in small quantity may also be added, if preferred.

SECTION C.—*Bread of the third kind.*

Of this there are several kinds. Those which are made by a simple effervescence, provided the residuum is not injurious, are best, and shall accordingly be placed first in order. Next will follow various kinds of bread made by the ordinary process of fermentation, salting, etc.

RECEIPT 1.—Wheat meal, seven pounds; carbonate of soda or saleratus[26] three quarters of an ounce to one ounce; water, two and three quarter pints; muriatic acid, 420 to 560 drops. Mix the soda with the meal as intimately as possible, by means of a wooden spoon or stick. Then mix the acid and water, and add it slowly to the mass, stirring it constantly. Make three loaves of it, and bake it in a quick oven.

RECEIPT 2.—Wheat meal, one pound; sesquicarbonate of soda, forty grains; muriatic acid, fifty drops; cold water, half a pint, or a sufficient quantity. Mix in the same way, and with the same caution, as in Receipt 1. Make one loaf of it, and bake in a quick oven.[27]

RECEIPT 3.—Wheat meal, one quart; cream of tartar, two tea-spoonfuls; saleratus, one tea-spoonful; and two and a half teacups full of milk. Mix well, and bake thirty minutes. If the meal is fresh, as it ought to be, the milk may be omitted.

RECEIPT 4.—Coarse rye meal, Indian meal, and oat meal, may be formed into bread in nearly a similar manner. So, in fact, may fine meal and all sorts of mixtures.

RECEIPT 5.—Professor Silliman more than intimates, that carbonic acid gas *might* be made to inflate bread, without either an effervescence or a fermentation. The plan is, to force carbonic acid, by some means or other, into the mass of dough, or, as bakers call it, the sponge. I do not know that the experiment has yet been made.

RECEIPT 6.—Coarse Indian meal may be formed into small, rather thin loaves, and prepared and baked as in Receipt 3.

Let us now proceed to common fermented bread:

RECEIPT 7.—Wheat meal, six pounds; good yeast, a teacup full; and a sufficient quantity of pure water. Knead thoroughly. Bake it in small loaves, unless you have a very strong heat.

RECEIPT 8.—Another way: Wheat meal, six quarts; molasses and yeast, each a teacup full. Mould into loaves half the thickness you mean they shall be after they are baked. Place them in the pans, in a temperature which will cause a moderate fermentation. When risen enough, place them in the oven. A strong heat is required.

RECEIPT 9.—Rye bread may be made in a similar way. It must, however, be well kneaded, to secure an intimate mixture with the yeast. Does not require quite so strong a heat as the former.

RECEIPT 10.—Oat meal bread may be prepared by mixing good kiln dried oat meal, a little salt and warm water, and a spoonful of yeast. Beat till it is quite smooth, and rather a thick batter; cover and let it stand to rise; then bake it on a hot iron plate, or on a bake stove. Be careful not to burn it.

RECEIPT 11.—Barley, or black bread, as it is called in Europe, makes a wholesome article of food. It may be fermented or unfermented.

RECEIPT 12.—Corn bread is sometimes made thus: Six pints meal, four pints water, one spoonful of salt; mix well, and bake in oblong rolls two inches thick. Bake in a hot oven.

It should be added to this division of my subject, that in baking bread sweet oil may be used (a vegetable oil) as a substitute for animal oil, to prevent the bread from adhering too closely. Or you may sift a quantity of Indian meal into the pans. If you use sweet, or olive oil, be sure to get that which is not rancid. Much of the olive oil of the shops is unfit to be used.

DIVISION II.—WHOLE GRAINS.

Some have maintained that since man is made to live on grain, fruits, etc., and since the most perfect mastication is secured by the use of uncooked grains, it is useless, and worse than useless, to resort to cookery at all, especially the cookery of bread. I have mentioned Dr. Schlemmer and his followers already as holding this opinion. Many of these people confine themselves to the use of uncooked grains and fruits. They do not cook their beans and peas. Nor can it be denied that they enjoy thus far very good health.

Now, while I admit that man, as an individual, can get along very well in this way, I am most fully persuaded that many kinds of farinaceous food are improved by cookery. Of the potato, I have already, incidentally, spoken. But are not wheat and corn, and many other grains, as well as the potato, improved by cookery? A barrel of flour (one hundred and ninety-six pounds) will make about two hundred and seventy pounds of good dry bread. It does not appear that the bread contains more water than the grain did from which it was made. Whence, then, the increase of weight by seventy-four pounds? Is not the water—a part of it, at least—which is used in making bread, rendered solid, as water is in slacking lime; or at least so incorporated with the flour or meal as to add both to its weight, and to its nutritious properties?

Or if, in the present infancy of the science of domestic chemistry, we are not able to give a satisfactory answer to the question, is not an affirmative highly probable? Such an answer would give no countenance, I believe, to the custom of raising our bread, since the increase of weight in making unfermented cakes or loaves, is about as great as in the case of fermented ones.

One of the strongest arguments ever yet brought against bread-making is, that it relieves us from the necessity of mastication. But to this we reply, that such cakes as may be made (and such loaves even) require more mastication than the uncooked grains. Pereira, in his excellent work on Diet, endeavors to support the doctrine that cooking bursts the grains of the farinacea, so as to bring them the better within the power of the stomach. This is specious, if not sound. In any event, I think it pretty certain, that though man can do very well on raw grains, yet there is a gain by cookery which more than repays the trouble. But though baking the flour or meal into cakes or bread, is the best method of preparation, there are other methods, secondary to this, which deserve our notice. One of these I will now describe.

SECTION A.—*Boiled Grains.*

These require less mastication than those which are submitted to other processes; but they are more easy of digestion, and to some more palatable, and even more digestible.

RECEIPT 1.—Take good perfect wheat; wash clean, and boil till soft in pure soft water. Those who are accustomed to salt their food, use sugar, etc., will naturally salt and sweeten this.

RECEIPT 2.—Rye or barley may be prepared in the same way, but it is not quite so sweet.

RECEIPT 3.—Indian corn may be boiled, but the process requires six hours or more, even after it has soaked all night, and there has been a frequent change of the water. And with all this boiling, the skins sometimes adhere rather strongly, unless you boil with them some ashes, or other alkali.

RECEIPT 4.—Rice, carefully cleaned, and well boiled, is good food. Imperfectly boiled, it is apt to disorder the bowels. And so unstimulating is it, and so purely nutritious, that they who eat it exclusively, without salt or curry, or any other condiment, are apt to become constipated. Potatoes go well with it.

RECEIPT 5.—Chestnuts, well selected, and well boiled, are highly palatable, greatly nutritious, and easy of digestion. They are best, however, soon after they are ripe.

RECEIPT 6.—Boiled peas, when ripe, either whole or split, make a healthy dish. They are best, however, when they have been cooked several days. When boiled enough, drain them through a sieve, but not very dry.

Some housekeepers soak ripe peas over night, in water in which they have dissolved a little saleratus. If you boil new or unripe peas, be careful not to cook them too much.

RECEIPT 7.—Beans, whether ripe or green (unless in bread or pudding), are not so wholesome as peas. They lead to flatulence, acidity, and other stomach disorders. And yet, eaten in moderate quantities, when ripe, they are to the hard, healthy laborer very tolerable food. Eaten green, they are most palatable, but least healthy.

RECEIPT 8.—Green corn boiled is bad food. Sweet corn, cooked in this way, is the best.

RECEIPT 9.—Lentils are nutritious, highly so; but I know little about them practically.

SECTION B.—*Grains, etc., in other forms. They may be baked, parched, roasted, or torrefied.*

RECEIPT 1.—Dry slowly, with a pretty strong heat, till they become so dry and brittle as to fall readily into powder. Corn is most frequently prepared in this way for food; but this and several other grains are often torrefied for coffee. Care should be taken to avoid burning.

RECEIPT 2.—Roasted grains are more wholesome. It is not usual or easy to roast them properly, however, except the chestnut, as the expanded air bursts or parches them. By cutting through the skin or shell, this result may be avoided, as it often is in the case of the chestnut. To roast well, they should be laid on the hearth or an iron plate, covered with ashes, and by building a fire slowly, all burning may be prevented.

RECEIPT 3.—Corn and buckwheat are often parched, and they form, especially the former, a very good food. In South America, and in some semi-barbarous nations, parched corn is a favorite dish.

RECEIPT 4.—Green corn is often roasted in the ear. It is less wholesome, however, than when boiled. Sweet corn is the best for either purpose.

RECEIPT 5.—Of baking grains I have little to say, because I *know* little on that subject.[28]

DIVISION III.—CAKES

This species of farinaceous food is much used, and is fast coming into vogue. The term, in its largest sense, would include the unleavened bread or cakes, of which I have spoken so freely in Division 1. They are for the most part, however, made by the addition of butter, eggs, aromatics, milk, etc., to the dough; and in proportion as they depart from simple bread, are more and more unhealthy. I shall mention but a few, though hundreds might be named which would still be vegetable food, as good olive oil, in preparing them, may be substituted for butter. I shall treat of them under one head or section.

RECEIPT 1.—Take of dough, prepared according to the English patented process, mentioned in Division I., Section C, Receipt 1 and Receipt 2, and bake in a thin form and in the usual manner.

RECEIPT 2.—Fruit cakes, if people will have them, may be made in the same manner. No butter would be necessary, even to butter eaters, when prepared in this patented way. If any have doubts, let them consult Pereira on Food and Diet, page 153.

RECEIPT 3.—Gingerbread may be made in the same way, and without alum or potash. It is thus comparatively harmless. Coarse meal always makes better gingerbread than fine flour.

RECEIPT 4.—Buckwheat cakes may be raised in the same general way.

RECEIPT 5.—Cakes of millet, rice, etc., are said to have been made by this process; but on this point I cannot speak from experience.

RECEIPT 6.—Biscuits, crackers, wafers, etc., are a species of cake, and might be made so as to be comparatively wholesome.

RECEIPT 7.—Biscuits may be made of coarse corn meal, with the addition of an egg and a little water. Make it into a stiff paste, and roll very thin.

DIVISION IV.—PUDDINGS.

These are a species of bread, only made thinner. They are usually unfermented. I shall speak of two kinds—hominy and puddings proper.

SECTION A.—*Hominy.*

This is usually eaten hot; but it improves on keeping a day or two. It may be warmed over, if necessary.

RECEIPT 1.—Wheat hominy, or cracked wheat, may be made into a species of pudding thus: Stir the hominy into boiling water (a little salted, if it must be so), very gradually. Boil from fifteen minutes to one hour. If boiled too long, it has a raw taste.

RECEIPT 2.—Corn hominy, or, as it is sometimes called, samp. Two quarts of hominy; four quarts of water; stir well, that the hulls may rise; then pour off the water through a sieve, that the hulls may separate. Pour the same water again upon the hominy, stir well, and pour off again several times.

Finally, pour back the water, add a little salt, if you use salt at all, and if necessary, a little more water, and hang it over a slow fire to boil. During the first hour it should be stirred almost constantly. Boil from three to six hours.

RECEIPT 3.—Another way: Take white Indian corn broken coarsely, put it over the fire with plenty of water, adding more boiling water as it wastes. It requires long boiling. Some boil it for six hours the day before it is wanted, and from four to six the next day. Salt, if used at all, may be added on the plate.

RECEIPT 4.—Another way still of making hominy is to soak it over night, and boil it slowly for four or five hours, in the same water, which should be soft.

There are other ways of making hominy, but I have no room to treat of them.

SECTION B.—*Puddings proper.*

These are of various kinds. Indeed, a single work I have before me on Vegetable Cookery has not less than 127 receipts for dishes of this sort, to say nothing of its pancakes, fritters, etc. I shall select a few of the best, and leave the rest.

The greatest objection to puddings is, that they are usually swallowed in large quantity, unmasticated, after we have eaten enough of something else. They are also eaten new and hot, and with butter, or some other mixture almost as injurious. Some puddings, from half a day to a day and a half old, are almost as good for us as bread.

One of the best puddings I know of, is a stale loaf of bread, steamed. Another is good sweet kiln dried oat meal, without any cooking at all. But there are some good cooked puddings, I say again, such as the following:

RECEIPT 1.—Boiled Indian pudding: Indian meal, a quart; water, a pint; molasses, a teacup full. Mix it well, and boil four hours.

RECEIPT 2.—Another Indian pudding. Indian meal, three pints; scald it, make it thin, and boil it about six hours.

RECEIPT 3.—Another of the same: To one quart of boiling milk, while boiling, add a teacup full of Indian meal; mix well, and add a little molasses. Boil three hours in a strong heat.

RECEIPT 4.—Hominy: Take a quart of milk and half a pint of Indian meal; mix it well, and add a pint and a half of cooked hominy. Bake well in a moderate oven.

RECEIPT 5.—Baked Indian pudding may be made by putting together and baking well a quart of milk, a pint of Indian meal, and a pint of water. Add salt or molasses, if you please.

RECEIPT 6.—Oat meal pudding: Pour a quart of boiling milk over a pint of the best fine oat meal; let it soak all night; next day add two beaten eggs; rub over, with pure sweet oil, a basin that will just hold it; cover it tight with a floured cloth, and boil it an hour and a half. When cold, slice and toast, or rather dry it, and eat it as you would oat cake itself.

This may be the proper place to say, that all coarse meal puddings are healthiest when twelve or twenty hours old; but are all improved—and so is brown bread—by drying, or almost toasting on the stove.

RECEIPT 7.—Rice pudding: To one quart of new milk add a teacup full of rice, sweetened a little. No dressings are necessary without you choose them. Bake it well.

RECEIPT 8.—Wheat meal pudding may be made by wetting the coarse meal with milk, and sweetening it a little with molasses. Bake in a moderate heat.

RECEIPT 9.—Boiled rice pudding may be made by boiling half a pound of rice in a moderate quantity of water, and adding, when tender, a coffee-cup full of milk, sweetening a little, and baking, or rather simmering half an hour. Add salt if you prefer it.

RECEIPT 10.—*Polenta*—Corn meal, mixed with cheese—grated, as I suppose, but we are not told in what proportion it is used—baked well, makes a pudding which the Italians call polenta. It is not very digestible.

RECEIPT 11.—Pudding may be made of any of the various kinds of meal I have mentioned, except those containing rye, by adding from one fourth to one third of the meal of the comfrey root. See Division I of this class, Section B, Receipt 17.

RECEIPT 12.—Bread pudding: Take a loaf of rather stale bread, cut a hole in it, add as much new milk as it will soak up through the opening, tie it up in a cloth, and boil it an hour.

RECEIPT 13.—Another of the same: Slice bread thinly, and put it in milk, with a little sweetening; add a little flour, and bake it an hour and a half.

RECEIPT 14.—Another still: Three pints of milk, one pound of baker's bread, four spoonfuls of sugar, and three of molasses. Cut the bread in slices; interpose a few raisins, if you choose, between each two slices, and then pour on the milk and sweetening. If baked, an hour and a half is sufficient. If boiled, two or three hours. Use a tin pudding boiler.

RECEIPT 15.—Rice and apple pudding: Boil six ounces of rice in a pint of milk, till it is soft; then fill a dish about half full of apples pared and cored; sweeten; put the rice over them as a crust, and bake it.

RECEIPT 16.—Stirabout is made in Scotland by stirring oat meal in boiling water till it becomes a thick pudding or porridge. This, with cakes of oat meal and potatoes, forms the principal food of many parts of Scotland.

RECEIPT 17.—Hasty pudding is best made as follows: Mix five or six spoonfuls of sifted meal in half a pint of cold water; stir it into a quart of water, while boiling; and from time to time sprinkle and stir in meal till it becomes thick enough. It should boil half or three quarters of an hour. It may be made of Indian or rye meal.

RECEIPT 18.—Potato pudding: Take two pounds of well boiled and well mashed potato, one pound of wheat meal; make a stiff paste, by mixing well; and tie it in a wet cloth dusted with flour. Boil it two hours.

RECEIPT 19.—Apple pudding may be made by alternating a layer of prepared apples with a layer of dough made of wheat meal, till you have filled a tin pudding boiler. Boil it three hours.

RECEIPT 20.—Sago pudding: Take half a pint of sago and a quart of milk. Boil half the milk, and pour it on the sago; let it stand half an hour; then add the remainder of the milk. Sweeten to your taste.

RECEIPT 21.—Tapioca pudding may be prepared in a similar manner.

RECEIPT 22.—To make cracker pudding, to a quart of milk add four thick large coarse meal crackers broken in pieces, a little sugar, and a little flour, and bake it one hour and thirty minutes.

RECEIPT 23.—Sweet apple pudding is made by cutting in pieces six sweet apples, and putting them and half a pint of Indian meal, with a little salt, into a pint of milk, and baking it about three hours.

RECEIPT 24.—Sunderland pudding is thus made: Take about two thirds of a good-sized teacup full of flour, three eggs, and a pint of milk. Bake about fifteen minutes in cups. Dress it as you please—sweet sauce is preferred.

RECEIPT 25.—Arrow root pudding may be made by adding two ounces of arrow root, previously well mixed with a little cold milk, to a pint of milk boiling hot. Set it on the fire; let it boil fifteen or twenty minutes, stirring it constantly. When cool, add three eggs and a little sugar, and bake it in a moderate oven.

RECEIPT 26.—Boiled arrow root pudding: Mix as before, only do not let it quite boil. Stir it briskly for some time, after putting it on the fire the second time, at a heat of not over 180 degrees. When cooled, add three eggs and a little salt.

RECEIPT 27.—Cottage pudding: Two pounds of potatoes, pared, boiled, and mashed, one pint of milk, three eggs, and two ounces of sugar, and if you choose, a little salt. Bake it three quarters of an hour.

RECEIPT 28.—Snow balls: Pare and core as many large apples as there are to be balls; wash some rice—about a large spoonful to an apple will be enough; boil it in a little water with a pinch of salt, and drain it. Spread it on cloths, put on the apples, and boil them an hour. Before they are turned out of the cloths, dip them into cold water.

Macaroni is made into puddings a great deal, and so is vermicelli; but they are at best very indifferent dishes. Those who live solely to eat may as well consult "Vegetable Cookery," where they will find indulgences enough and too many, even though flesh and fish are wholly excluded. They will find soups, pancakes, omelets, fritters, jellies, sauces, pies, puddings, dumplings, tarts, preserves, salads, cheese-cakes, custards, creams, buns, flummery, pickles, syrups, sherbets, and I know not what. You will find them by hundreds. And you will find directions, too, for preparing almost every vegetable production of both hemispheres. And if you have brains of your own you may invent a thousand new dishes every day for a long time without exhausting the vegetable kingdom.

DIVISION V.—PIES.

Pies, as commonly made, are vile compounds. The crust is usually the worst part. The famous Peter Parley (S. G. Goodrich, Esq.), in his Fireside Education, represents pies, cakes, and sweetmeats as totally unfit for the young.

Within a few years attempts have been made to get rid of the crust of pies— the abominations of the crust, I mean—by using Indian meal sifted into the pans, etc.; but the plan has not succeeded. It is the pastry that gives pies their charm. Divest them of this, and people will almost as readily accept of plain ripe fruit, especially when baked, stewed, or in some other way cooked.

As pies are thus objectionable, and are, withal, a mongrel race, partaking of the nature both of bread and fruit, and yet, as such, unfit for the company of either, I will almost omit them. I will only mention two or three.

RECEIPT 1.—Squashes, boiled, mashed, strained, and mixed with milk or milk and water, in small quantity, may be made into a tolerable pie. They may rest on a thick layer of Indian meal.

RECEIPT 2.—Pumpkins may be made into pies in a similar manner; but in general they are not so sweet as squashes.

RECEIPT 3.—Potato pie: Cut potatoes into squares, with one or two turnips sliced; add milk or cream, just to cover them; salt a little, and cover them with a bread crust. Sweet potatoes make far better pies than any other kind.

Almost any thing may be made into pies. Plain apple pies—so plain as to become mere apple sauce—are far from being very objectionable. See the next Class of Foods.

CLASS II.—FRUITS.

So far as fruits, at least in an uncooked state, have been used as food, they have chiefly been regarded as a dessert, or at most as a condiment. Until within a few years, few regarded them as a principal article—as standing next to bread in point of importance. In treating of these substances as food, I shall simply divide them into Domestic and Foreign.

DIVISION I.—DOMESTIC FRUITS.

SECTION A.—*The large fruits—Apple, Pear, Peach, Quince, etc.*

RECEIPT 1.—The apple. May be baked in tin pans, or in a common bake pan. The sweet apple requires a more intense heat than the sour. The skin may be removed before baking, but it is better to have it remain. The best apple pie in the world is a baked apple.

RECEIPT 2.—It may be roasted before the fire, by being buried in ashes, or by throwing it upon hot coals, and quickly turning it. The last process is sometimes called *hunting* it.

RECEIPT 3.—It may be boiled, either in water alone, or in water and sugar, or in water and molasses. In this case the skin is often removed, that the saccharine matter may the better penetrate the body of the apple.

RECEIPT 4.—It may also be pared and cored, and then stewed, either alone or with molasses, to form plain apple sauce—a comparatively healthy dish.

RECEIPT 5.—Lastly, it may be pared and cored, placed in a deep vessel, covered with a plain crust, as wheat meal formed into dough, and baked slowly. This forms a species of pie.

RECEIPT 6.—The pear is not, in every instance, improved by cookery. Several species, however, are fit for nothing, till mid-winter, when they are either boiled, baked, or stewed.

The peach can hardly be cooked to advantage. It is sometimes cut up, and sprinkled with sugar and other substances.

RECEIPT 7.—A tolerably pleasant sauce can be made by stewing or baking the quince, and adding sugar or molasses, but it is not very wholesome.

SECTION B.—*The smaller fruits. The Strawberry, Cherry, Raspberry, Currant, Whortleberry, Mulberry, Blackberry, Bilberry, etc.*

None of these, so far as I know, are improved by cookery. It is common to stew green currants, to make jams, preserves, sauces, etc., but this is all wrong. The great Creator has, in this instance, at least, done his own work, without leaving any thing for man to do.

There is one general law in regard to fruits, and especially these smaller fruits. Those which melt and dissolve most easily in the mouth, and leave no residuum, are the most healthy; while those which do not easily dissolve—which contain large seeds, tough or stringy portions, or hulls, or scales—are in the same degree indigestible.

I have said that fruits were next to bread in point of importance. They are to be taken, always, as part of our regular meals, and never between meals. Nor should they be eaten at the end of a meal, but either in the middle or at the beginning. And finally, they should be taken either at breakfast or dinner. According to the old adage, fruit is gold in the morning, silver at noon, and lead at night.

DIVISION II.—FOREIGN FRUITS.

The more important of these are the banana, pine-apple, and orange, and fig, raisin, prune, and date. The first three need no cooking, two of the last four may be cooked. The date is one of the best—the orange one of the worst, because procured while green, and also because it is stringy.

RECEIPT 1.—The prune. Few things sit easier on the feeble or delicate stomach than the stewed prune. It should be stewed slowly, in very little water.

RECEIPT 2.—The good raisin is almost as much improved by stewing as the prune.

I do not know that the fig has ever yet been subjected to the processes of modern cookery. It is, however, with bread, a good article of food.

Fruits, in their juices, may be regarded as the milk of adults and old people, but are less useful to young children and to the *very* old. But to be useful they must be perfectly ripe, and eaten in their season. Thus used, they prevent a world of summer diseases—used improperly, they invite disease, and do much other mischief.

In general, fruits and milk do not go very well together. The baked sweet apple and whortleberry seem to be least objectionable.

CLASS III.—ROOTS.

DIVISION I.—MEALY ROOTS.

These are the potato, in its numerous varieties, the artichoke, the groundnut, and the comfrey. Of these the potato is by far the most important.

SECTION A.—*The Common Potato.*

This may be roasted, baked, boiled, steamed, or fried. It is also made into puddings and pies. Roasting in the ashes is the best method of cooking it; frying by far the worst. I take this opportunity to enter my protest against all frying of food. Com. Nicholson, of revolutionary memory, would never, as his daughters inform me, have a frying-pan in his house.

The potato is best when well roasted in the ashes, but also excellent when baked, and very tolerable when boiled or steamed.

There are many ways of preparing the potato and cooking it. Some always pare it. It may be well to pare it late in the winter and in the spring, but not at other times. For, in paring, we lose a portion of the richest part of the potato, as in the case of paring the apple. There is much tact required to pare a potato properly, that is, thinly.

RECEIPT 1.—To boil a potato, see that the kettle is clean, the water pure and soft, and the potatoes clean. Put them in as soon as the water boils.[29] When they are soft, which can be determined by piercing them with a fork, pour off the water, and let them steam about five minutes.

RECEIPT 2.—To roast in the ashes, wash them clean, then dry them, then remove the heated embers and ashes quite to the bottom of the fire-place, and place them as closely together as possible, but not on top of each other. Cover as quickly as possible, and fill the crevices with hot embers and small coals. Let them be as nearly of a size as possible, and cover them to the depth of an inch. Then build a hot fire over them. They will be cooked in from half an hour to three quarters of an hour, according to the size and heat of the fire.

RECEIPT 3.—Baking potatoes in a stove or oven, is a process so generally known, that it hardly needs description.

RECEIPT 4.—Steaming is better than boiling. Some fry them; others stew them with vegetables for soup, etc.

SECTION B.—*The Sweet Potato.*

This was once confined to the Southern States, but it is now raised in tolerable perfection in New Jersey and on Long Island. It is richer than the common potato in saccharine matter, and probably more nutritious; but not, it is believed, quite so wholesome. Still it is a good article of food.

RECEIPT 1.—Roasting is the best process of cooking these. They may be prepared in the ashes or before a fire. The last process is most common. They cook in far less time than a common potato.

RECEIPT 2.—Baking and roasting by the fire are nearly or quite the same thing as respects the sweet potato. Steaming is a little different, and boiling greatly so. The boiled sweet potato is, however, a most excellent article.

DIVISION II.—SWEET AND WATERY ROOTS.

These are far less healthy than the mealy ones; and yet are valuable, because, like potatoes, they furnish the system with a good deal of innutritious matter, to be set off against the almost pure nutriment of bread, rice, beans, peas, etc.

RECEIPT 1.—The beet is best when boiled thoroughly, which requires some care and a good deal of time. It may be roasted, baked, or stewed, however. It is rich in sugar, but is not very easily digested.

RECEIPT 2.—The parsnep. The boiled parsnep is more easily *dissolved* in the stomach than the beet; but my readers must know that many things which are dissolved in the stomach are nevertheless very imperfectly digested.

RECEIPT 3.—The turnip, well boiled, is watery, but easily digested and wholesome. It may also be roasted or baked, and some eat it raw.

RECEIPT 4.—The carrot is richer than the turnip, but not therefore more digestible. It may be boiled, stewed, fried, or made into pies, puddings, etc. It is a very tolerable article of food.

RECEIPT 5.—The radish, fashionable as it is, is nearly useless.

RECEIPT 6.—For the sick, and even for others, arrow root jellies, puddings, etc., are much valued. This, with sago, tapioca, etc., is most useful for that class of sick persons who have strong appetites.[30]

CLASS IV.—MISCELLANEOUS ARTICLES OF FOOD.

Under this head I shall treat briefly of the proper use of a few substances commonly and very properly used as food, but which cannot well come under any of the foregoing classes. They are chiefly found in the various chapters of my Young Housekeeper, as well as in Dr. Pereira's work on Food and Diet, under the heads of "Buds and Young Shoots," "Leaves and Leaf Stalks," "Cucurbitaceous Fruits," and "Oily Seeds."

RECEIPT 1.—Asparagus, well boiled, is nutritious and wholesome. Salt is often added, and sometimes butter. The former, to many, is needless; the latter, to all, injurious.

RECEIPT 2.—Some of the varieties of the squash are nutritious and wholesome, especially when boiled. Its use in pies and puddings is also well known.

RECEIPT 3.—A few varieties of the pumpkin, especially the sweet pumpkin, are proper for the table. Made into plain sauce, they are highly valued by most, but they are best known as ingredients of pies and puddings. A few eat them when merely baked.

RECEIPT 4.—The tomato is fashionable, but a sour apple, if equal pains were taken with it, and it were equally fashionable, might be equally useful. It adds, however, to nature's vast variety!

RECEIPT 5.—Watermelons, coming as they do at the end of the hot season, when eaten with bread, are happily adapted (as most other ripe fruits are, when eaten in the same way, and at their own proper season) to prevent disease, and promote health and happiness.

RECEIPT 6.—Muskmelons are richer than watermelons, but not more wholesome. Of the canteloupe I know but little.

RECEIPT 7.—The cucumber. Taken at the moment when ripe—neither green nor acid—the cucumber is almost, but not quite as valuable as the melon. It should be eaten in the same way, rejecting the rind. The Orientals of modern days sometimes boil them, but in former times they ate them uncooked, though always ripe. Unripe cucumbers are a *modern* dish, and will erelong go out of fashion.

RECEIPT 8.—Onions have medicinal properties, but this should be no recommendation to healthy people. Raw, they are unwholesome; boiled, they are better; fried, they are positively pernicious.

RECEIPT 9.—Nuts are said to be adapted to man in a state of nature; but I write for those who are in an artificial state, not a natural state. Of the chestnut I have spoken elsewhere. The hazelnut is next best, then perhaps the peanut and the beechnut. The butternut, and walnut or hickory-nut, are too oily. Nor do I see how they can be improved by cookery.

RECEIPT 10.—Cabbage, properly boiled, and without condiments, is tolerable, but rather stringy, and of course rather indigestible.

RECEIPT 11.—Greens and salads are stringy and indigestible. Besides, they are much used, as condiments are, to excite or provoke an appetite—a thing usually wrong. A feeble appetite, say at the opening of the spring, however common, is a great blessing. If let alone, nature will erelong set to rights those things, which have gone wrong perhaps all winter; and then appetite will return in a natural way.

But the worst thing about greens, salads, and some other things, is, they are eaten with vinegar. Vinegar and all substances, I must again say, which resist or retard putrefaction, retard also the work of digestion. It is a universal law, and ought to be known as such, that whatever tends to preserve our food—except perhaps ice and the air-pump—tends also to interfere with the great work of digestion. Hence, all pickling, salting, boiling down, sweetening, etc., are objectionable. Pereira says, "By drying, salting, smoking, and pickling, the digestibility of fish is greatly impaired;" and this, except as regards *drying*, is but the common doctrine. It should, however, be applied generally as well as to fish.

FOOTNOTES:

[25] Formerly called Graham meal.

[26] I shall use these terms indiscriminately, as they mean in practice the same thing.

[27] Both these processes are patented in Great Britain. The bread thus retains its sweetness—no waste of its saccharine matter, and no residuum except muriate of soda or common salt. Sesquicarbonate of soda is made of three parts or atoms of the carbonic acid, and two of the soda.

[28] Keep butter and all greasy substances away from every preparation of food which belongs to this division—especially from green peas, beans, corn, etc.

[29] Some prepare them, and soak them in water over the night.

[30] In general, the appetites of the sick are taken away by design. In such cases there should be none of the usual forms of indulgence. A little bread—the crust is best—is the most proper indulgence. If, however, the appetite is raging, as in a convalescent state it sometimes is, puddings and even gruel may be proper, because they busy the stomach without giving it any considerable return for its labor.